A FUNCTORIAL MODEL THEORY

Newer Applications to Algebraic Topology, Descriptive Sets, and Computing Categories Topos

A FUNCTORIAL MODEL THEORY

Newer Applications to Algebraic Topology, Descriptive Sets, and Computing Categories Topos

Cyrus F. Nourani

Apple Academic Press

TORONTO NEW JERSEY

Apple Academic Press Inc.
3333 Mistwell Crescent
Oakville, ON L6L 0A2
Canada

Apple Academic Press Inc.
9 Spinnaker Way
Waretown, NJ 08758
USA

Printed in the United States of America on acid-free paper

International Standard Book Number-13: 978-1-926895-92-5 (Hardcover)

Library of Congress Control Number: 2013955135

Library and Archives Canada Cataloguing in Publication

Nourani, Cyrus F., author
A functorial model theory : newer applications to algebraic topology, descriptive sets, and computing categories topos/Cyrus F. Nourani.

Includes bibliographical references and index.
ISBN 978-1-926895-92-5
1. Functor theory. 2. Model theory. I. Title.

QA169.N69 2014 512'.62 C2013-907778-2

ABOUT THE AUTHOR

Cyrus F. Nourani

Dr. Cyrus F. Nourani has a national and international reputation in computer science, artificial intelligence, mathematics, virtual haptic computation, information technology, and management. He has many years of experience in the design and implementation of computing systems. Dr. Nourani's academic experience includes faculty positions at the University of Michigan-Ann Arbor, the University of Pennsylvania, the University of Southern California, UCLA, MIT, and the University of California, Santa Barbara. He was also a Research Professor at Simon Frasier University in Burnaby, British Columbia, Canada. He was a Visiting Professor at Edith Cowan University, Perth, Australia, and a Lecturer of Management Science and IT at the University of Auckland, New Zealand.

Dr. Nourani commenced his university degrees at MIT where he became interested in algebraic semantics. That was pursued with a category theorist at the University of California during a brief time to completion. Dr. Nourani's dissertation on computing models and categories proved to have intuitionists forcing developments that were published from the postdoctoral times on the Association for Symbolic Logic. He has taught AI to the Los Angeles aerospace industry and has authored many R&D and commercial ventures. He has written and co-authored several books. He has over 350 publications in mathematics and computer science and has written on additional topics, such as pure mathematics, AI, EC, and IT management science, decision trees, predictive economics game modeling. In 1987, he founded ventures for computing R&D. He began independent consulting with clients such as System Development Corporation (SDC), the US Air Force Space Division, and GE Aerospace. Dr. Nourani has designed and developed AI robot planning and reasoning systems at Northrop Research and Technology Center, Palos Verdes, California. He also has comparable AI, software, and computing foundations and R&D experience at GTE Research Labs.

CONTENTS

PREFACE

This book is a preliminary introduction to a functorial model theory based on infinitary language categories. The perspective is different from the preceding authors', in that functorial model theory is based on defining categories on language fragments, then carrying on functors to sets and categories to develop models. Infinitary language categories are defined, and their preliminary categorical properties are presented. A foundation for infinitary language categories is presented to avoid having to define a categorical interpretation for logic, for example, Heyting algebra. An important consequence is that we have a direct reach to all developed model theory techniques at categories. Therefore, we do not reinvent all that at topos, but carry that at topos. Thus, the present mathematics opens new areas that can be explored across models to topos. The author has defined Infinitary Language Categories, abbreviated IFLCs, and has shown their preliminary properties in 1994–1995 brief abstracts. In this trend, a model theory is defined for functors starting with a countable fragment of an infinitary language. The infinite language category defined on Lω1, K generic sets on the Lω1, K, the respective Kiesler fragment and a functors from Lω1, K to Set, called generic functors are defined. A new technique for generating generic models with categories is defined by inventing infinite language categories and functorial model theory. Specific functor models, the String Models, are defined by infinite chains on fragment models on an infinite language category functor.

Techniques similar to Robinson's consistency theorem are invented to define limit models. Functorial models are further defined on a generalized diagram by limit chains. The techniques are further developed over the past several years to essentially filling the gap between forcing and toposes. The techniques are further developed to mount untraproducts and ultrafilters, compressed onto categories on algebraic theories, to reach for newer model products on categories, onto homotopy with models on categories. Projectives on categories are reached ahead of hindsight to a chapter in which projective set models are applied with generic filters to create or address saturated models at categories. Fragment consistent algebras are developed as new techniques that are applied to realizability and higher stratified consistency or completeness. Positive omitting types and forcing techniques on topos were gaps in our mathematics at categories that are in part developed here in the chapters and abstracts published over a decade.

Functorial models, fragment consistent algebras, prime and saturated models are developed to study product fields with reduced or ultraproduct generic filter-based techniques to present new product field factorization areas with functors on product fields. Horn filters ultraproducts on projective sets towards presheaves are newer areas treated, allowing us to address models, algebraic set theory, and set axioms. Reliability, computability, and models for product languages are the further application areas. Topos and algebraic topology are developed with the newer techniques to apply Stone representation with newer product topologies defined here on fragment models to have a new perspective on the Lowenhiem-Skolem-Tarski questions on infinite cardinal models, stating topos counterparts for infinite cardinal models.

The book concludes with a glimpse on algebraic set theory where the author presents Skolem hull models for sets on descriptive functors, projective sets, and saturated models. Newer areas are merged with what the author had written on generic filter functors. Horn product models on set foundations and the recent focus on set axioms are presented with functorial models, mapping models for Godel sets onto Boolean models. The concluding chapter opens to a basis for arbitrary functorial models based on descriptive infinitary languages, where the book's techniques are applicable for newer developments. The book is written for a first- or second-year graduate courses in mathematics with exercises that at times are good starts to a dissertation area.

— Cyrus F. Nourani
Berlin, November 28, 2013.

CHAPTER 1

INTRODUCTION

CONTENTS

The book starts with an introduction to categories from a model theoretic perspective. Categorical axiomatizations (Carnap, 1943), categoricity, and functors direct from syntax that were bases for intuitionistic models are the author's gentle introduction towards topoi characterizations for categorical models and infinite language categories. Nearly all chapters, except for Chapter 2, are recent original areas developed, partly attributed to the author, as a new functorial model theory that reaches to areas on forcing on topos that were only on the author's abstracts over a decade.

Hilbert like axiom of completeness that ascribe to the objects of an axiomatic theory a maximality property, in that they assert there is no comprehensive system of objects that also satisfies a series on axioms, might be called a maximal axiom (c.f. Chapter 10). The same axiomatic role as that of maximal axioms is played in minimal axioms, which ascribe a minimal property to the elements of the discipline. In monomorphic closed axiom systems, maximal axiom systems are equivalent with the positive requirement that every model of the axioms can be isomorphically mapped into M.

There are three important themes in the categorical approach to logic:

Categorical semantics with categorical logic introduces the notion of structure valued in a category C with the classical model theoretic notion of a structure appearing in the particular case where C is the category of sets and functions. This notion has proven useful when the set-theoretic notion of a model lacks generality and/or is inconvenient. R.A.G. Seely's modeling of various impredicative theories, such as system F is an example of the usefulness of categorical semantics internal languages. This can be seen as a formalization and generalization of proof by diagram chasing. One defines a suitable internal language naming relevant constituents of a category, and then applies categorical semantics to turn assertions in logic over the internal language into corresponding categorical statements.

This has been most successful in the theory of toposes, where the internal language of a topos together with the semantics of intuitionistic higher-order logic in a topos enables one to reason about the objects and morphisms of a topos "as if they were sets and functions". This has been successful in dealing with toposes that have "sets" with properties incompatible with classical logic. A prime example is Dana Scott's model of untyped lambda calculus in terms of objects that retract onto their own function space. Another is the Moggi-Hyland model of system F by an internal full subcategory of the effective topos of Martin Hyland.

1.1 CATEGORIES AND MODELS

In many cases, the categorical semantics of logic provide a basis for establishing a correspondence between theories in the logic and instances of an appropriate kind of category. A classic example is the correspondence between theories of βη-equational logic over simply typed lambda calculus and Cartesian closed categories. Categories arising from theories via term-model constructions can usually be characterized up to equivalence by a suitable universal property. This has enabled proofs of meta-theoretical properties of some logics by means of an appropriate categorical algebra. For instance, Freyd gave a proof of the existence and disjunction properties of intuitionistic logic this way.

Categorical logic became a mathematical discipline when Lawvere's *Functorial Semantics of Algebraic Theories* (1963) presented a basis to address categories on theories onto and *Elementary Theory of the Category of Sets* (1964). Lawvere recognized the Grothendieck topos, introduced in algebraic topology as a generalized space, as a generalization of the category of sets, on mathematical logic on quantifiers and sheaves (1970). With Myles Tierney, Lawvere then developed the notion of elementary topos, thus establishing the fruitful field of topos theory, which provides a unified categorical treatment of the syntax and semantics of higher-order predicate logic. The resulting logic is formally intuitionistic. Andre Joyal is credited, in the term Kripke–Joyal semantics, with the observation that the sheaf models for predicate logic, provided by topos theory, generalize Kripke semantics. Joyal and Moerdijk (1995) applied these models to study higher-order concepts such as the real numbers in the intuitionistic setting.

An analogous development was the link between the simply typed lambda calculus and Cartesian-closed categories (Lawvere, 1963; Lambek and Scott, 1988), which provided a setting for the development of domain theory. Less expressive theories, from the mathematical logic viewpoint, have their own category theory counterparts. For example the concept of an algebraic theory leads to Gabriel–Ulmer duality. The view of categories as a generalization unifying syntax and semantics has been productive in the study of logics and type theories for applications in computer science.

The founders of elementary topos theory were Lawvere and Tierney. Lawvere's writings, sometimes couched in a philosophical jargon, isolated some of the basic concepts as adjoint functors (which he explained as 'objective' in a Hegelian sense, not without some justification). A subobject classifier is a strong property to ask of a category, since with Cartesian closure and finite limits it gives a topos (axiom bashing shows how strong the assumption is). Lawvere's further work in the 1960 s gave a theory of attributes, which in a sense is a subobject theory more in sympathy with type theory. Major influences subsequently have been Martin-Löf type theory from the direction of logic, type polymorphism and the calculus of constructions from functional programming, linear logic from proof theory, game semantics and the projected synthetic domain theory. The abstract categorical idea of fibration has been much applied.

Intuitionistic logic had reappeared in mathematics, in a central place in the Bourbaki–Grothendieck program, a generation after the messy Brouwer–Hilbert controversy had ended, with Hilbert the apparent winner. Bourbaki, or more accurately Jean Dieudonné, having laid claim to the legacy of Hilbert and the Göttingen school including Emmy Noether, had revived intuitionistic logic's credibility although Dieudonné himself found intuitionistic logic ludicrous, as the logic of an arbitrary topos, where classical logic was that of the topos of sets. This was one consequence, certainly unanticipated, of Grothendieck's view; and not lost on Pierre Cartier, one of the broadest of the core group of French mathematicians around Bourbaki and IHES. Cartier was to give Bourbaki exposition of intuitionistic logic.

The reliability concept of Kleene is also a serious route to intuitionistic logic towards categorical logic. From Kolmogorov there is one way to explain the protean Curry–Howard isomorphism. The Curry-Howard-Lambek correspondence provides a deep isomorphism between intuitionistic logic, simply typed lambda calculus and Cartesian closed categories. Newer insights are on Schiemer (2012), whereas this author's beginning were based on a dissertation that prove to be on intuitionistic forcing (Nourani, 1983) and a functorial metamathematics (Nourani, 2006) to establish a basis to areas, with the manuscript is essentially new insights on the areas treating the gap from forcing on toposes over a decade. Certain

Cartesian closed categories, the topoi, have been proposed as a general setting for mathematics, instead of traditional set theory. The renowned computer scientist John Backus has advocated a variable-free notation, or function-level programming, which in retrospect bears some similarity to the internal language of Cartesian closed categories. CAML is more consciously modeled on Cartesian closed categories

In mathematics and logic, a **higher-order logic** is a form of predicate logic that is distinguished from first-order logic by additional quantifiers and a stronger semantics. Higher-order logics with their standard semantics are more expressive, but their model-theoretic properties are less well behaved than those of first-order logic.

The term "higher-order logic", abbreviated as **HOL**, is commonly used to mean **higher order simple predicate logic**. Here "simple" indicates that the underlying type theory is simple, not polymorphic or dependent.

There are two possible semantics for HOL. In the **standard** or **full semantics**, quantifiers over higher-type objects range over all possible objects of that type. For example, a quantifier over sets of individuals' ranges over the entire power set of the set of individuals. Thus, in standard semantics, once the set of individuals is specified, this is enough to specify all the quantifiers.

HOL with standard semantics is more expressive than first-order logic. For example, HOL admits categorical axiomatizations of the natural numbers, and of the real numbers, which are impossible with first-order logic. However, by a result of Gödel, HOL with standard semantics does not admit an effective, sound and complete proof calculus.

The model-theoretic properties of HOL with standard semantics are also more complex than those of first-order logic. For example, the Löwenheim number of second-order logic is already larger than the first measurable cardinal, if such a cardinal exists. The Löwenheim number of first-order logic, in contrast, is $\aleph_0$, the smallest infinite cardinal.

In Henkin semantics, a separate domain is included in each interpretation for each higher-order type. Thus, for example, quantifiers over sets of individuals may range over only a subset of the power set of the set of individuals. HOL with these semantics is equivalent to many-sorted first-order logic, rather than being stronger than first-order logic. In particular, HOL with Henkin semantics has all the model-theoretic properties of first-order logic, and has a complete, sound, effective proof system inherited from first-order logic. Examples of higher order logics include HOL, Church's *Simple Theory of Types*. The model theory is couched in terms of a specific infinitary fragment categories as far as the language base is. But since we are stating the theorems on sets and topos, the generalizations to arbitrary infinitary languages are apparent. For example, Barwise L-admissible infinitary logics are implicit on certain chapters.

The exposition begins on Chapter 2, which addresses the categorical and mathematical preliminaries on models. Categories, functors, morphisms are presented followed by categorical products. The basic concepts of limits and colimits, cones, and functor categories are presented following the standard presentations, for example, MacLane. Adjoint functors are introduced with the basic theorems on adjoint functors, followed by set and poset categories. Universal maps, natural transformations, and representations are briefly shown with examples and Yoneda lemma. Limits and preservation areas are examined. From that point on topos and model theory are introduced based on the original Lawever and Grothendiek characterizations. Glimpses on algebraic topology appear in a paragraph or two but are treated in two to three chapters, from Chapter 5 on. Monoidal categories are introduced observing that any category is a generalized monoid. Heyting algebras are an awakening to intuitionist logic at algebras starters. More universal constructions are observed towards Lindebaum algebras and Cartesian closed categories.

Chapter 3 introduces a first presentation on the author's 1995 infinite language categories (Nourani, 1995–2007): a category direct on "Logisch Syntax der Sprache" since Carnap. Categories are defined on infinitary language fragments based on the Kiesler's omitting type bases to infinitary model theory. Starting with small complete fragment language categories techniques for creating generic models with generic functors are presented where functors create fragment models based on language fragments: termed Functorial String Models (Nourani, 2005). Functors on morphic structures are defined capable to characterize Skolem hulls on end extension models. That topic is treated in Chapter 4 further.

Categorical grammars are treated next with a functorial linguistics comparing to early accomplishments by Lambek from a newer functorial treatment. Versatile abstract syntax categories are examined followed with new computational examples from infinite product languages with Kleene initial algebraic models.

1.2 FRAGMENT CONSISTENT KLEENE MODELS

Chapter 4 starts with developing a pure functorial model theory for categorical models with fragment consistent models with newer infinitary language product consistency models and towers on infinite chain structures. Product models, embedding, and elementary generic fragment diagrams are applied to create functorial models. New stratified consistency and completeness is presented with fragment bases. Horn product models, positive morphism and positive categories are introduced with new techniques to create models with generic functors creating limit models with infinitary consistency piecemeal model generation. Fragment positive omitting types algebras and positive realizably bases are introduced on positive fragment models from the author's positive omitting types and realizability publications (Nourani, 1983–2007, 2012), (Van den Berg, 2012) and c.f. Realizability Workshop 2012, Savoie, France for newer group accomplishments.

Chapter 5 starts with products on models towards topos characterizations for algebraic theories and models. Basic categorical and functorial algebraic theory models are presented in light of ultraproducts on models embedding and structural model products. Ultraproducts and ultrapowers from Kiesler et al. (1964) have a brief introduction with further specifics in Chapter 10 for set model areas. Basic reduced product models on Horn categories are presented from this author's ASL brief over the past recent years. Ultraproducts on algebras are presented from a new perspective toward the direct algebraic theory product categories Lawever style. Starting from a direct algebraic theoretic basis from Lawvere this chapter defines topos models based on free theories, congruence, factor theories, and T-algebras. Adjunctions on algebraic theory categories, limits, colimits, and cones are stated for algebraic theory categories with functor theories, theory morphisms, products and coproducts. T-algebras and projections are presented to compare to reduce product constructions for further insights. Natural transformations are presented for the above T-categories.

Chapter 6 starts with elementary topos where we begin to apply generic functors. Initial categories on generic diagram models and positive forcing are applied to characterizing models on topos, vending fragment models from limit cones. Chapter 3 and here (Chapter 6) is where the basics for forcing on topos are presented.

Topos on models from sheaves to Stone representations and structures on topologies, models on topos, are introduced to conclude with a glimpse on standard models with cardinality exposures, for example, *Löwenheim-Skolem Theorems*. Skolem is reputed not to believe the upward direction of the theorem himself since he did not "believe" the existence of uncountable sets. Nonetheless that is an important area to address. That might make our readers feel better that we can have a characterization with on topos on that area.

Positive forcing models, based on the author's publications of over a decade are presented towards specific techniques for forcing applications to topos addressing gaps in such mathematics areas. Example model computing on topos with Hasse diagrams are presented. Next, fragment consistent models from the preceding chapters are revisited with a specific direction towards topos, starting with filters on presheaves on a fragment topology with generic functors. Filters on categories and filter categories are both addressed, the former a new area, with applications to the latter older known areas. Homotopy theories on topos are stated since Artin and Mazur composed this with the forgetful functor a contravariant functor. Brown's representability theorem in algebraic topology is reviewed on Elinberg Maclane Space. Carrying into new areas on types and models, the chapter develops a new functorial characterization for Martin Lof of type theory based on infinite language generic functor models on omitting type topos, comparing to the past decade's accomplishments on models for categories from authors. The chapter concludes with filtered colimits towards comma categories

Chapter 7 is on models, sheaves, and topos. It starts with presheaves duality, fragment models and topological structures on fragments. Foundation areas on algebraic topology is briefed towards the Stone (1934) representation theorem—a fundamental area. The Scott topology is introduced by way of comparing categoricity directions on topological structures. Newer lifts on topos models on cardinalities are presented towards a topos characterization for Löwenheim-Skolem theorems. This is a new area the author had stated on a proposition over a year ago (Nourani, 2012) but not all topological structures were developed at that time to accomplish the goals due to lack of time assigned to students and newer alumni that did not carry on the areas. So the author included that for this new work.

Filters on categories are further studied for the saturation on higher cardinality models. Newer product topological structures with adjoint functors on saturation cardinalities are specific new developments to accomplish. Chapter 8 develops a new area on functorial fields. Basic field models are presented based on generic functors on infinite language categories and Vaught's prime model theorem. Galois fields are addressed on algebraically closed fields. Omitting types on fields are applied with the new positive omitting type functors from Chapters 3 and 4, to present field models and filters on fields. Complete theories, Horn product models, and generic functors are presented with new applications on algebraically closed fields. Fragment consistent models allow us to reach to a new product field factorization theorem with important applications to algebraically closed fields.

Chapter 9 further develops generic functors on fragment product models and completing theories on fragments. Prime model product completions on types realizations at cardinalities are example areas for completing theories on fragments. Generic products, filters, and ultraproducts are applied with projective sets to present projective set models on determinacy saturations. Saturation and preservation theorems are studied and generic filters on Horn models are presented. More on countable incomplete ultrafilters and ultraproduct and ultrafilters here carries onto topological structures for topos models on saturations. Positive categories and F-generic filer functors from the authors (2007) are presented. The book concludes with Chapter 10 with a glimpse on algebraic set theory—a new interest area in mathematics foundations. Filters, utraproducs and ultrapowers on saturation cardinals are further studied there.

Ultraproducts and ultrafilters on sets are examined in light of the preceding developments in the book. Ultrafilters over N, saturation and preservation on positive fragments ultraproducts are briefed. Functorial models and descriptive sets that were developed from 2007 by the authors based on admissible fragments are presented with specific applications to model end extensions and a new area for admissible hulls.

Rudimentary fragment models on descriptive set functors from the authors ASL briefs are stated. Newer statements generic filter sets were carried on by (Blass 2010).

Functorial fragment consistency on filters is examined further with application to Horn product models to Rasiowa-Sikorski lemma to present a Horn dense characterization for that important foundations lemma. Forcing axioms, MA Maximus, based on filters, fragment constructible models, and sets based on the newer foundations are reviewed with newer accomplishments. Functors on \\V and Boolean models with the Gödel universe is presented for a direct reach to an algebraic set theory. All theorems, lemmas, and propositions from Chapter 3 on that are not explicitly credited to preceding authors are either well known or due to the present author. On occasions, to establish intellectual direction on the minds, theorems are explicitly annotated with this author's name.

KEYWORDS

- **higher-order logic**
- **Löwenheim-Skolem Theorems**
- **quantifiers**
- **sheaves**
- **theory of toposes**

CHAPTER 2

CATEGORICAL PRELIMINARIES

CONTENTS

2.1 CATEGORIES AND FUNCTORS

A *category C* is comprised of:

1. A class **C,** with elements called objects;
2. A class denoted *M* with elements called *morphisms*. Each morphism ***f*** has a *source object* ***a*** and *target object* ***b***. The expression $f : a \to b$, would be verbally stated as "*f* is a morphism from *a* to *b*". The expression **hom(*a*, *b*)** — alternatively expressed as **hom*C*(*a*, *b*)**, **mor(*a*, *b*)**, or ***C*(*a*, *b*)** — denotes the *hom-class* of all morphisms from *a* to *b*.
3. A binary operation ∘, called *composition of morphisms*, such that for any three objects *a*, *b*, and *c*, we have hom(*b*, *c*) × hom(*a*, *b*) → hom(*a*, *c*). The composition of $f : a \to b$ and $g : b \to c$ is written as $g^{\circ}f$ or *gf*, [3] governed by two axioms:
   a. Associativity: If $f : a \to b$, $g : b \to c$ and $h: c \to d$ then $h\circ (g\circ f) = (h\circ g) \circ f$, and
   b. Identity: For every object x, there exists a morphism $1x : x \to x$ called the *identity morphism* for *x*, such that for every morphism $f : a \to b$, we have $1b\circ f = f = f \circ 1a$.

From the axioms, it can be proved that there is exactly one identity morphism for every object. Some authors deviate from the definition just given by identifying each object with its identity morphism.

I. Objects will generally be denoted with capital letters A, B, C in **C.**
II. For morphisms we write f: A→ B or A →f→ B to mean that f has source A and target B. We say that "f is a morphism from A to B."
III. *M* (A, B) denotes the class of all morphisms from A to B, and is called the hom-set of *M.*

More succinct is the following:

A *category C* consists of the following three mathematical entities:

1. A class ob(*C*), whose elements are called *objects*;
2. A class hom(*C*), whose elements are called morphisms or maps or *arrows*. Each morphism ***f*** has a *source object* ***a*** and *target object* ***b***. The expression $f : a \to b$, would be verbally stated as "*f* is a morphism from *a* to *b*". The expression hom(*a*, *b*) — alternatively expressed as hom*C*(*a*, *b*), mor(*a*, *b*), or *C*(*a*, *b*) — denotes the *hom-class* of all morphisms from *a* to *b*.

3. A binary operation $\circ$, called *composition of morphisms*, such that for any three objects *a*, *b*, and *c*, we have hom(*b*, *c*) × hom(*a*, *b*) → hom(*a*, *c*). The composition of $f : a \to b$ and $g : b \to c$ is written as $g \circ f$ or *gf*, governed by two axioms:
   a. Associativity: If $f : a \to b$, $g : b \to c$ and $h : c \to d$ then $h \circ (g \circ f) = (h \circ g) \circ f$, and
   b. Identity: For every object *x*, there exists a morphism $1x : x \to x$ called the *identity morphism for x*, such that for every morphism $f : a \to b$, we have $1b \circ f = f = f \circ 1a$.

From the axioms, it can be proved that there is exactly one identity morphism for every object. Some authors deviate from the definition just given by identifying each object with its identity morphism.

2.2 MORPHISMS

Relations among morphisms (such as fg = h) are often depicted using commutative diagrams, with "points" (corners) representing objects and "arrows" representing morphisms.

Morphisms can have any of the following properties. A morphism f : a → b is a:

- *monomorphism* (or *monic*) if $f \circ g_1 = f \circ g_2$ implies $g_1 = g_2$ for all morphisms $g_1, g_2 : x \to a$.
- *epimorphism* (or *epic*) if $g_1 \circ f = g_2 \circ f$ implies $g_1 = g_2$ for all morphisms $g_1, g_2 : b \to x$.
- *bimorphism* if *f* is both epic and monic.
- *isomorphism* if there exists a morphism $g : b \to a$ such that $f \circ g = 1b$ and $g \circ f = 1a$.
- *endomorphism* if $a = b$. end(*a*) denotes the class of endomorphisms of *a*.
- *automorphism* if *f* is both an endomorphism and an isomorphism. aut(*a*) denotes the class of automorphisms of *a*.
- *retraction* if a right inverse of *f* exists, that is, if there exists a morphism $g : b \to a$ with $fg = 1b$.
- *section* if a left inverse of *f* exists, that is, if there exists a morphism $g : b \to a$ with $gf = 1a$.

Every retraction is an epimorphism, and every section is a monomorphism. Furthermore, the following three statements are equivalent:

- f is a monomorphism and a retraction;
- f is an epimorphism and a section;
- f is an isomorphism.

2.3 FUNCTORS

Functors are structure-preserving maps between categories. They can be thought of as morphisms in the category of all (small) categories.

A *(covariant)* functor F from a category C to a category D, written $f: C \rightarrow D$, consists of:

- for each object x in C, an object $F(x)$ in D; and
- for each morphism $f: x \rightarrow y$ in C, a morphism $F(f): F(x) \rightarrow F(y)$,

such that the following two properties hold:

- For every object x in C, $F(1x) = 1F(x)$;
- For all morphisms $f: x \rightarrow y$ and $g: y \rightarrow z$, $F(g \circ f) = F(g) \circ F(f)$.

A *contravariant* functor $f: C \rightarrow D$, is like a covariant functor, except that it "turns morphisms around" ("reverses all the arrows"). More specifically, every morphism $f: x \rightarrow y$ in C must be assigned to a morphism $F(f)$: $F(y) \rightarrow F(x)$ in D. In other words, a contravariant functor acts as a covariant functor from the opposite category $C^{op} \rightarrow D$.

Using the language of category theory, many areas of mathematical study can be categorized. Categories include sets, groups, topologies, and so on.

Each category is distinguished by properties that all its objects have in common, such as the empty set or the product of two topologies, yet in the definition of a category, objects are considered to be atomic, that is, we do not know whether an object A is a set, a topology, or any other abstract concept. Hence, the challenge is to define special objects without referring to the internal structure of those objects. To define the empty set without referring to elements, or the product topology without referring to open sets, one can characterize these objects in terms of their relations to other objects, as given by the morphisms of the respective categories. Thus, the task is to find *universal properties* that uniquely determine the objects of interest.

Indeed, it turns out that numerous important constructions can be described in a purely categorical way. The central concept, which is needed for this purpose, is called categorical *limit*, and can be dualized to yield the notion of a *colimit.*

Definition 2.1 Suppose $S : D \rightarrow C$ is a functor and c an object of C: a universal arrow from c to S is a pair <r, u> consisting of an object r of D and an arrow u : $c \rightarrow$ Sr of C, such that for every paid <d, f> with d an object r od D and $f : c \rightarrow$ Sd and arrow of C there is a unique arrow f' : r $\rightarrow$ d of D with sf'$_{0}$ $u = f$. That is, every arrow f to S factors uniquely through the universal arrow u, as the commutative diagram.

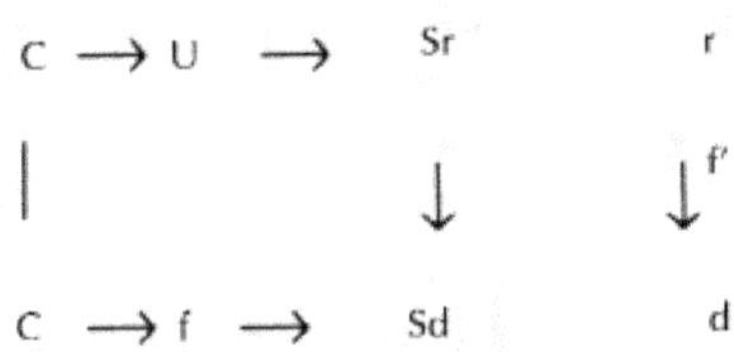

Definition 2.2 Suppose that U: $D \rightarrow C$ is a functor from a category D to a category C, and let X be an object of C. Consider the following dual (opposite) notions:

- An *initial morphism* from X to U is an initial object in the category (X ¯ U) of morphisms from X to U. In other words, it consists of a pair (A, φ) where A is an object of D and φ: $X \rightarrow U(A)$ is a morphism in C, such that the following initial property is satisfied:
  Whenever Y is an object of D and $f : X \rightarrow U(Y)$ is a morphism in C, then there exists a *unique* morphism $g : A \rightarrow Y$ such that the following diagram commutes:
- A *terminal morphism* from U to X is a terminal object in the comma category (U ¯ X) of morphisms from U to X. In other words, it consists of a pair (A, φ) where A is an object of D and φ: $U(A) \rightarrow X$ is a morphism in C, such that the following *terminal property* is satisfied:

Whenever Y is an object of D and f: $U(Y) \rightarrow X$ is a morphism in C, then there exists a *unique* morphism $g : Y \rightarrow A$ such that the following diagram commutes:

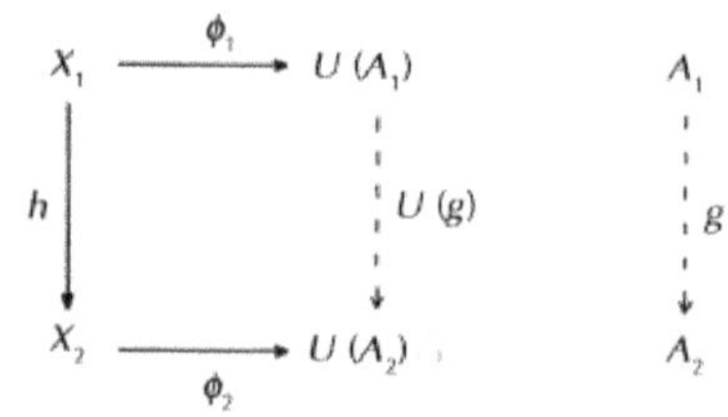

The term *universal morphism* refers either to an initial morphism or a terminal morphism, and the term universal property refers either to an initial property or a terminal property. In each definition, the existence of the morphism g intuitively expresses the fact that (A, φ) is "general enough", while the uniqueness of the morphism ensures that (A, φ) is "not too general". In most cases the function embedding a mathematical object in a suitably completed object can be interpreted as a universal arrow. The general fact on uniqueness of the universal arrow implies the uniqueness of the completed object, up to isomorphisms. An object is initial in a category **C** if to each object a there is exactly one arrow $s \rightarrow a$. Universality in terms of a universal element is as follows. Consider the equivalence relation E on a set S, the corresponding quotient set S/E. S/E has the property that any function f on S.

S ---p⟶ S/E

h↓ ↓f′

S′ ---f ⟶ X

which respect the equivalence relation can be regarded as a function on S/E.

$f : S \to X$ has $fs = fs$,' whenever $s\ E\ s'$ then f can be written as a composite $f = f'p$, for a unique function f': S/E → X, with the following commutative diagram:

That is, <S/E, p> is a universal element if the functor H : **Set** → **Set** *which assigns to each set X the set HX of all the functions* $f : S \to X$ *for* which $s\ E\ s'$ implies $fs = fs'$. The notions of *initial* and *terminal* are dual, it is often enough to discuss only one of them, and simply reverse arrows in C for the dual discussion. Alternatively, the word *universal* is often used in place of both words.

Note: some authors may call only one of these constructions a *universal morphism* and the other one a *couniversal morphism*.

2.4 CATEGORICAL PRODUCTS

A categorical product can be characterized by a terminal property. For concreteness, one may consider the Cartesian product in **Set**, the direct product in **Grp**, or the product topology in **Top**.

Let X and Y be objects of a category D. The product of X and Y is an object $X \times Y$ together with two morphisms:

- $\pi_1 : X \times Y \to X$
- $\pi_2 : X \times Y \to Y$

such that for any other object Z of D and morphisms $f : Z \to X$ and $g : Z \to Y$ there exists a unique morphism $h : Z \to X \times Y$ such that $f = \pi_1 \circ h$ and $g = \pi_2 \circ h$.

To understand this characterization as a terminal property we take the category C to be the product category $D \times D$ and define the diagonal functor,

$$\Delta : D \to D \times D$$

by $\Delta(X) = (X, X)$ *and* $\Delta(f : X \to Y) = (f, f)$. *Then* $(X \times Y, (\pi 1, \pi 2))$ *is a* terminal morphism from Δ to the object (X, Y) of $D \times D$. This is just a above restatement, since the pair (f, g) represents an (arbitrary) morphism from $\Delta(Z)$ to (X, Y).

2.4.1 LIMITS AND COLIMITS

Categorical products are a particular kind of limit in category theory. One can generalize the above example to arbitrary limits and colimits.

Let J and C be categories with J a small index category and let CJ be the corresponding functor category. The *diagonal functor* $\Delta : C \rightarrow CJ$ is the functor that maps each object N in C to the constant functor $\Delta(N)$: $J \rightarrow C$ to N (i.e., $\Delta(N)(X) = N$ for each X in J).

Given a functor $f : J \rightarrow C$ (thought of as an object in CJ), the *limit* of F, if it exists, is nothing but a terminal morphism from Δ to F. Dually, the *colimit* of F is an initial morphism from F to Δ.

Defining a quantity does not guarantee its existence. Given a functor U and an object X as above, there may or may not exist an initial morphism from X to U. If, however, an initial morphism (A, φ) does exist then it is essentially unique. Specifically, it is unique up to a *unique* isomorphism: if (A', φ') is another such pair, then there exists a unique isomorphism $k : A \rightarrow A'$ such that $\varphi' = U(k)\varphi$. This is easily seen by substituting (A', φ') for (Y, f) in the definition of the initial property. It is the pair (A, φ), which is essentially unique in this fashion. The object A itself is only unique up to isomorphism. Indeed, if (A, φ) is an initial morphism and $k : A : A'$ is any isomorphism then the pair (A', φ'), where $\varphi' = U(k)\varphi$, is also an initial morphism.

The definition of a universal morphism can be rephrased in a variety of ways. Let U be a functor from D to C, and let X be an object of C. Then the following statements are equivalent:

- (A, φ) is an initial morphism from X to U
- (A, φ) is an initial object of the comma category $(X \downarrow U)$
- (A, φ) is a representation of Hom$C(X, U—)$

The dual statements are also equivalent:

- (A, φ) is a terminal morphism from U to X
- (A, φ) is a terminal object of the comma category $(U \downarrow X)$
- (A, φ) is a representation of Hom$C\,(U—, X)$

2.4.2 ADJOINT FUNCTORS

An adjunction between categories C and D is a pair of functors, $\mathrm{F} :\ \mathrm{D} \rightarrow \mathrm{C}$ and $g : \mathrm{C} \rightarrow \mathrm{D}$ and a family of bijections

$\hom_C(FY, X) \cong \hom_D(Y, GX)$

which is natural in the variables *X* and *Y*. The functor *F* is called a *left adjoint functor*, while *G* is called a *right adjoint functor*. The relationship "*F* is left adjoint to *G*" (or equivalently, "*G* is right adjoint to *F*") is sometimes written F |— G.

Suppose (A_1, φ_1) is an initial morphism from X_1 to *U* and (A_2, φ_2) is an initial morphism from X_2 to *U*. By the initial property, given any morphism $h: X_1 \to X_2$ there exists a unique morphism $g : A_1 \to A_2$ such that the following diagram commutes:

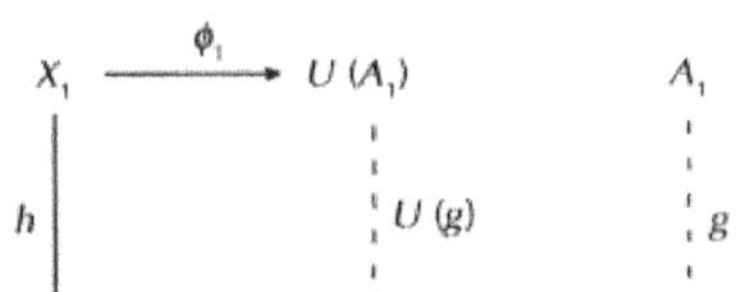

If *every* object *Xi* of *C* admits an initial morphism to *U*, then the assignment Xi a Ai and *h* a *g* defines a functor V from *C* to *D*. The maps φi then define a natural transformation from 1*C* (the identity functor on *C*) to *UV*. The functors (*V*, *U*) are then a pair of adjoint functors, with *V* left-adjoint to *U* and *U* right-adjoint to *V*.

Similar statements apply to the dual situation of terminal morphisms from *U*. If such morphisms exist for every *X* in *C* one obtains a functor *V*: $C \to D$ which is right-adjoint to *U* (so *U* is left-adjoint to *V*).

Indeed, all pairs of adjoint functors arise from universal constructions in this manner. Let *F* and *G* be a pair of adjoint functors with unit η and co-unit ε (see the article on adjoint functors for the definitions). Then we have a universal morphism for each object in *C* and *D*:

For each object *X* in *C*, $(F(X), \eta X)$ is an initial morphism from *X* to *G*. That is, for all $f : X \to G(Y)$ there exists a unique $g : F(X) \to Y$ for which the following diagrams commute.

For each object *Y* in *D*, $(G(Y), \varepsilon Y)$ is a terminal morphism from *F* to *Y*. That is, for all $g : F(X) \to Y$ there exists a unique $f : X \to G(Y)$ for which the following diagrams commute. Universal constructions are more general than adjoint functor pairs: a universal construction is like an optimization problem; it gives rise to an adjoint pair if and only if this problem has a solution for every object of C (equivalently, every object of D).

Definition 2.3 Let E be a field extension of a field F. Given a set of elements A in the larger field E we denote by $F(A)$ the smallest subextension, which contains the elements of A. We say $F(A)$ is constructed by *adjunction* of the elements A to F or *generated* by A.

If A is finite we say $F(A)$ is **finitely generated** and if A consists of a single element we say $F(A)$ is a **simple extension**. The primitive element theorem states a finite separable extension is simple. In a sense, a finitely generated extension is a transcendental generalization of a finite extension since, if the generators in A are all algebraic, then $F(A)$ is a finite extension of F. Because of this, most examples come from algebraic geometry. A subextension of a finitely generated field extension is also a finitely generated extension.

Notes: $F(A)$ consists of all those elements of E that can be constructed using a finite number of field operations +, –, *, / applied to elements from F and A. For this reason $F(A)$ is sometimes called the *field of rational expressions* in F and A.

For example, given a field extension E/F then $F(\emptyset) = F$ and $F(E) = E$.

The complex numbers are constructed by adjunction of the imaginary unit to the real numbers, that is $\mathbf{C}=\mathbf{R}\,(i)$.

Given a field extension E/F and a subset A of E, let t be the family of all finite subsets of A. Then $F(A) = \cup\, F(T)$

$T \in t$

In other words, the adjunction of any set can be reduced to a union of adjunctions of finite sets. Given a field extension E/F and two subset N, M of E then $K(M \cup N) = K(M)(N) = K(N)(M)$. This shows that any adjunction of a finite set can be reduced to a successive adjunction of single elements.

In category theory, a branch of mathematics, the functors between two given categories form a category, where the objects are the functors and the morphisms are natural transformations between the functors. Functor categories are of interest for two main reasons:

2.4.3 POSETS AND SETS

Definition 2.4 Let **C** be a locally small category and let **Set** be the category of sets. For each object A of **C** let Hom(A, –) be the hom functor which maps objects X to the set Hom(A, X).

A functor f : **C** → **Set** is said to be *representable* if it is naturally isomorphic to Hom(*A*, –) for some object *A* of **C**. A *representation* of *F* is a pair (*A*, Φ) where

Φ: Hom(*A*, –) → *F* is a natural isomorphism.

A contravariant functor *G* from **C** to **Set** is the same thing as a functor *g* : $\mathbf{C}^{op}$ → **Set** and is therefore representable just when it is naturally isomorphic to the contravariant hom-functor Hom(–, *A*) for some object *A* of **C**.

A *universal element* of a functor *f* : **C** → **Set** is a pair (*A*, *u*) consisting of an object *A* of **C** and an element $u \in F(A)$ such that for every pair (*X*, *v*) with $v \in F(X)$ there exists a unique morphism $f: A \rightarrow X$ such that $(Ff)u = v$.

A universal element may be viewed as a universal morphism from the one-point set {•} to the functor *F* or as an initial object in the category of elements of *F*.

The natural transformation induced by an element u ∈ *F*(*A*) is an isomorphism if and only if (*A*, *u*) is a universal element of *F*. We therefore conclude that representations of *F* are in one-to-one correspondence with universal elements of *F*. For this reason, it is common to refer to universal elements (*A*, *u*) as representations.

Consider the contravariant functor P : **Set** → **Set** which maps each set to its power set and each function to its inverse image map. To represent this functor we need a pair (*A*, *u*) where *A* is a set and u is a subset of *A*, that is, an element of *P*(*A*), such that for all sets *X*, the hom-set Hom(*X*, *A*) is isomorphic to *P*(*X*) via $\Phi X(f) = (Pf)u = f^{-1}(u)$. Take *A* = {0,1} and u = {1}. Given a subset S ⊆ X the corresponding function from X to A is the characteristic function of S.

Forgetful functors to **Set** are very often representable. In particular, a forgetful functor is represented by (*A*, *u*) whenever *A* is a free object over a singleton set with generator u. The forgetful functor **Grp** → **Set** on the category of groups is represented by (**Z**, 1).

The forgetful functor **Ring** → **Set** on the category of rings is represented by (**Z**[x], x), the polynomial rings in one variable with integer coefficients.

The forgetful functor **Vect** → **Set** on the category of real vector spaces is represented by (**R**, 1). The forgetful functor **Top** → **Set** on the category of topological spaces is represented by any singleton topological space with its unique element.

A group G can be considered a category (even a groupoid) with one object that we denote by •. A functor from *G* to **Set** then corresponds to a *G*-set. The unique hom-functor Hom(•, –) from *G* to **Set** corresponds to the canonical *G*-set *G* with the action of left multiplication. Standard arguments from group theory show that a functor from *G* to **Set** is representable if and only if the corresponding *G*-set is simply transitive (i.e., a *G*-torsor). Choosing a representation amounts to choosing an identity for the group structure.

Definition 2.5 *A relation R on a set A is a pair* ⇒ *of maps* $p_1, p_2: R \rightarrow A$. *Given* $R\ (p_1, p_2) \Rightarrow A$ *let the equivalence relation generated by (R, p1, p2) be ≡ R, the smallest equivalence relation containing the relation defined by R:* $(a_1, a') \in \equiv R$ *iff either a=a' or (a, a') can be linked by a chain* $(a_1, a_2, \ldots, a_n+1)$ *of elements of A when* $a=a_1$, *a'=an+1 and for each intermediate* $k \leq n$ *either* (a_k, a_k+1) *or* $(a_n+1, an) \in \equiv R$.

Lemma 2.1 *If* $R\ (p_1, p_2) \Rightarrow$ *generates* $==_R$ *then the diagram* $R\ (p_1, p_2) \Rightarrow A \rightarrow h \rightarrow A/\equiv_R$ *has the property that if h' satisfies* $R\ (p_1, p_2) \Rightarrow A \rightarrow h' \rightarrow B$ *then there is a unique* $\Psi: A/\equiv_R \rightarrow B$ *such that*

R (p1,p2) ⇒ A → A/ ≡ R

↘ ↙Ψ

B

2.4.3 REPRESENTATION AND UNIVERSAL ELEMENT

A representable functor is a functor of a special form from an arbitrary category into the category of sets. Such functors give representations of an abstract category in terms of known structures (i.e., sets and functions) allowing one to use, as much as possible, knowledge about the category of sets in other settings.

From another point of view, representable functors for a category *C* are the functors *given* with *C*. Their theory is a vast generalization of upper sets in posets, and of Cayley's theorem in group theory.

Representations of functors are unique up to a unique isomorphism. That is, if (A_1, Φ_1) and (A_2, Φ_2) represent the same functor, then there exists a unique isomorphism φ: $A_1 \rightarrow A_2$ such that

$$\Phi_{1-}^{\ 1} \circ \Phi_2 = (\varphi_{,} —)$$

as natural isomorphisms from Hom(A_2, –) to Hom(A_1, –). This fact follows easily from Yoneda's lemma.

Stated in terms of universal elements: if (A_1, u_1) and (A_2, u_2) represent the same functor, then there exists a unique isomorphism $\varphi : A_1 \rightarrow A_2$ such that

$$(F\varphi)u_1, u_2.$$

To encounter a representation functor, consider the author's 1994 Tours ECCT (Nourani, 1994) presentation, let us start with a preliminary form to the adjoint functor theorem.

We start with the following preliminary form to Fryed's adjoint functor.

Theorem 2.1 (Freyd) Let D be a small-complete category with small hom-sets. Then D has an initial object if and only if it satisfies the following called the Solution Set Condition. There exists a small set I and an I-indexed family ki of objects of D such that for every d in D there is an i in I and arrow ki –> d of D.

The solution set conditions are necessary for presenting initial objects and form the basis for the above preliminary theorem and to what is referred to as the adjoint functor theorem and many basic results in category theory: if D has an initial object k, then k indexed by the one-point set realizes the condition, since always a unique arrow. Let us consider some definitions from the author's 1994 papers that allow us to define standard models for of theories by abstract recursion. We present techniques for obtaining solution sets by defining a model theory with a countable fragment of an infinitary logic for categories. It is shown how categories defined on an infinitary language have limits from which solution sets are definable at an adjoining category. Further directions for research are presented as a consequence. Examining some definitions from the author's views to standard models of theories. The standard models are significant for tree computational theories that we had presented and the intelligent tree computation theories developed by the present paper.

Definition 2.6 *Let M be a set and F a finite family of functions on M. We say that (F, M) is a monomorphic pair, provided for every f in F, f is injective, and the sets {Image(f):f in F} and M are pair-wise disjoint.*

This definition is basic in defining induction for abstract recursion-theoretic hierarchies and inductive definitions. We define generalized standard models with monomorphic pairs.

Definition 2.7 A standard model M, of base M and functionality F, is a structure inductively defined by <F, M> provided the <F, M> forms a monomorphic pair.

Now we define generic model-diagrams for Initial Models. The generic diagram is a model diagram in which the elements of the structure are all represented by a minimal set of function symbols and constants, such that it is sufficient to define the truth of formulas only for the terms generated by the minimal set of functions and constant symbols. Such assignment implicitly defines the diagram. This allows us to define a canonical model of a theory in terms of a minimal family of function symbols. The minimal sets of functions that define a generic diagram are those with which a monomorphic pair could define a standard model. Σ1 Skolem functions are example diagram functions.

The following is to be referred to by the Solution Set Theorem.

Theorem 2.2 Generic diagrams and standard models defined by monomorphic terms from define solution set conditions for the adjoint functors, thus instantiating initial models and objects.

Proof Outline (more specifics on forthcoming chapters): Applies the above definitions and the glimpse on the natural correspondence between generic trees defining representatives term models to the arrow kià D where the small set I and an I-indexed family where every d in D there is an i in I and arrow ki → d of D. The arrow picks the generic term. The natural transformation is with the category of diagram models D<A, G>: objects are models and arrows are diagram model morphisms.

Alternative characterization for above on the solution set theorem from generic diagrams are on the following chapters, for example, Chapters 7 and 9.

2.5 NATURAL TRANSFORMATIONS

Many commonly occurring categories are (disguised) functor categories, so any statement proved for general functor categories is widely applicable; every category embeds in a functor category (via the Yoneda embedding); the functor category often has nicer properties than the original category, allowing certain operations that were not available in the original setting.

Definition 2.9 Suppose C is a small category (i.e., the objects and morphisms form a set rather than a proper class) and D is an arbitrary category. The category of functors from C to D, written as Fun(C, D), Funct(C, D) or DC, has as objects the covariant functors from C to D, and as morphisms the natural transformations between such functors. Note that natural transformations can be composed: if μ(X): F(X) → G(X) is a natural transformation from the functor F: C → D to the functor g : C → D, and η(X): G(X) → H(X) is a natural transformation from the functor G to the functor H, then the collection η(X) μ(X): F(X) → H(X) defines a natural transformation from F to H. With this composition of natural transformations (known as vertical composition, see natural transformation), DC satisfies the axioms of a category.

Analogously, one can also consider the category of all *contravariant* functors from *C* to *D*; we write this as Funct(C^{op}, → *D*). If *C* and *D* are both preadditive categories (i.e., their morphism sets are abelian groups and the composition of morphisms is bilinear), then we can consider the category of all additive functors from *C* to *D*, denoted by Add(*C*, *D*).

If I is a small discrete category, that is, its only morphisms are the identity morphisms, then a functor from I to C essentially consists of a family of objects of C, indexed by I; the functor category CI can be identified with the corresponding product category: its elements are families of objects in C and its morphisms are families of morphisms in C. An arrow category $\mathbf{C}^{\rightarrow}$ (whose objects are the morphisms of C, and whose morphisms are commuting squares in **C**, is just $\mathbf{C}^2$, where "2" is the category with two objects and their identity morphisms as well as an arrow from one object to the other, but not another arrow back the other way.

A directed graph consists of a set of arrows and a set of vertices, and two functions from the arrow set to the vertex set, specifying each arrow's start and end vertex. The category of all directed graphs is thus nothing but the functor category **Set** C, where C is the category with two objects connected by two morphisms, and **Set** denotes the category of sets. Any group G can be considered as a one-object category in which every morphism is invertible. The category of all G-sets is the same as the functor category **Set** G. Similar to the previous example, the category of k-linear representations of the group G is the same as the functor category ***k*-Vect** G (where ***k*-Vect** denotes the category of all vector spaces over the field k).

Any ring of left modules over R is the same as the additive functor category Add(R, **Ab**) (where **Ab** denotes the category of abelian groups), and the category of right R-modules is Add(R^{op}, **Ab**). Because of this example, for any preadditive category C, the category Add(C, **Ab**) is sometimes called the "category of left modules over C" and Add(C^{op}, **Ab**) is the category of right modules over C.

The category of presheaves on a topological space X is a functor category: we turn the topological space into a category C having the open sets in X as objects and a single morphism from U to V if and only if U is contained in V. The category of presheaves of sets (abelian groups, rings) on X is then the same as the category of contravariant functors from C to **Set** (or **Ab** or **Ring**). Because of this example, the category Funct(C^{op}, **Set**) is sometimes called the "category of presheaves of sets on C" even for general categories C not arising from a topological space. To define sheaves on a general category C, one needs more structure: a Grothendieck topology on C. Some authors refer to categories that are equivalent to **Set** C as presheaf categories.

Most constructions that can be carried out in *D* can also be carried out in *DC* by performing them "component wise", separately for each object in *C*. For instance, if any two objects X and *Y* in *D* have a product *X*×*Y*, then any two functors *F* and *G* in *DC* have a product *F*×*G*, defined by (F×G)(c) = F(c)×G(c) for every object c in C. Similarly, if ηc: *F*(c)→G(c) is a natural transformation and each ηc has a kernel Kc in the category D, then the kernel of η in the functor category *DC* is the functor K with *K*(c) = *Kc* for every object *c* in *C*.

As a consequence we have the general rule of thumb that the functor category *DC* shares most of the "nice" properties of *D*:

- if *D* is complete (or cocomplete), then so is *DC*;
- if *D* is an abelian category, then so is *DC*;

We also have: if *C* is any small category, then the category **Set** *C* of presheaves is a topos.

So from the above examples, we can conclude right away that the categories of directed graphs, *G*-sets and presheaves on a topological space are all complete and co-complete topoi, and that the categories of representations of *G*, modules over the ring *R*, and presheaves of abelian groups on a topological space *X* are all abelian, complete and co-complete.

A basic category lemma on representations is Yoneda Lemma. It can be stated in terms of an object c of a locally small category *C*, meaning one having a homfunctor C(−, −): Cop × *C* → Set (i.e., small homsets), and a functor f : *C* → Set or presheaf.

Lemma 2.2 (Yoneda). The function α : Hom(*C*(*c*, −), *F*) → *F*(*c*) defined by α (τ: C(c, −) → F) = τc(1c) is a bijection natural in c.

According to Yoneda's lemma, natural transformations from Hom(A, –) to F are in one-to-one correspondence with the elements of F(A). Given a natural transformation Φ: Hom(A, –) → F, the corresponding element u ∈ F(A) is given by $u = F_A(id_A)$.

Conversely, given any element u ∈ F(A) we may define a natural transformation

Φ: Hom(A, –) → F via $\Phi_X(f) = (Ff)(u)$.

where f is an element of Hom(*A*, *X*). In order to get a representation of F we want to know when the natural transformation induced by u is an isomorphism. This leads to the following definition:

The embedding of the category C in a functor category that was mentioned earlier uses the Yoneda lemma as its main tool. For every object X of C, let Hom(–, X) be the contravariant representable functor from C to **Set**. The Yoneda lemma states that the assignment X a Hom(¾, X) is a full embedding of the category *C* into the category Funct(C^{op}, **Set**). So *C* naturally sits inside a topos. The same can be carried out for any preadditive category *C*: Yoneda then yields a full embedding of C into the functor category Add(C^{op}, **Ab**). So C naturally sits inside an abelian category.

The intuition mentioned above (that constructions that can be carried out in *D* can be "lifted" to *DC*) can be made precise in several ways; the most succinct formulation uses the language of adjoint functors. Every functor f : D → *E* induces a functor *FC* : *DC* → *EC* (by composition with F). If *F* and *G* is a pair of adjoint functors, then *FC* and *GC* is also a pair of adjoint functors.

The functor category *DC* has all the formal properties of an exponential object; in particular the functors from *E* × *C* → *D* stand in a natural one-to-one correspondence with the functors from *E* to *DC*. The category **Cat** of all small categories with functors as morphisms is therefore a cartesian closed category.

2.6 PRESERVATION OF LIMITS

Representable functors are naturally isomorphic to Hom functors and therefore share their properties. In particular, (covariant) representable functors preserve all limits. It follows that any functor, which fails to preserve some limit is not representable. Contravariant representable functors take co-limits to limits.

Any functor K: C → **Set** with a left adjoint f : **Set** → C is represented by (FX, ηX(•)) where X = {•} is a singleton set and η is the unit of the adjunction.

Conversely, if K is represented by a pair (A, u) and all small co-powers of A exist in C then K has a left adjoint F which sends each set I to the I-th co-power of A.

Therefore, if C is a category with all small co-powers, a functor K: C → **Set** is representable if and only if it has a left adjoint.

The categorical notions of universal morphisms and adjoint functors can both be expressed using representable functors.

Let g : D → C be a functor and let X be an object of C. Then (A, φ) is a universal morphism from X to G if and only if (A, φ) is a representation of the functor HomC(X, G–) from D to **Set**. It follows that G has a left-adjoint F if and only if HomC(X, G–) is representable for all X in C. The natural isomorphism ΦX : HomD(FX, –) → HomC(X, G–) yields the adjointness; that is, $\Phi_{X, Y:}\mathrm{Hom}_D(FX, Y) \to \mathrm{Hom}_C(X, GY)$ is a bijection for all X and Y.

The dual statements are also true. Let f : C → D be a functor and let Y be an object of D. Then (A, φ) is a universal morphism from F to Y if and only if (A, φ) is a representation of the functor HomD(F–, Y) from C to **Set**. It follows that F has a right-adjoint G if and only if HomD (F–, Y) is representable for all Y in D.

Proposition 2.1 If f : A → B then the map k : A/≡f → f(A): $[a] \equiv_f \to f(a)$ is a bijection.

Proof (Exercise)

Let us start from certain model-theoretic premises with propositions known form basic model theory. A **structure** consists of a set along with a collection of finitary functions and relations, which are defined on it. Universal algebra studies structures that generalize the algebraic structures such as groups, rings, fields, vector spaces and lattices. Model theory has a different scope than universal algebra, encompassing more arbitrary theories, including foundational structures such as models of set theory. From the model-theoretic point of view, structures are the objects used to define the semantics of first-order logic. A **structure** can be defined as a triple consisting of a **domain** A, a signature Σ, and an **interpretation function** I that indicates how the signature is to be interpreted on the domain. To indicate that a structure has a particular signature σ one can refer to it as a σ-structure. The domain of a structure is an arbitrary set (nonempty); it is also called the **underlying set** of the structure, its **carrier** (especially in universal algebra), or its **universe** (especially in model theory). Sometimes the notation or is used for the domain of, but often no notational distinction is made between a structure and its domain, that is, same symbol can refer to both. The signature of a structure consists of a set of **function symbols** and **relation symbols** along with a function that ascribes to each symbol s a natural number, called the **arity** of s. For example, let G be a graph consisting of two vertices connected by an edge, and let H be the graph consisting of the same vertices but no edges. H is a subgraph of G, but not an induced substructure. Given two structures and of the same signature Σ', a **Σ-homomorphism** is a map that can preserve the functions and relations. A Σ-homomorphism h is called a Σ-embedding if it is one-to-one and for every n-ary relation symbol R of Σ and any elements $a_1, \ldots, a_n$, the following equivalence holds:

$R(a_1, \ldots, a_n)$ iff $R(h(a_1), \ldots h(a_n))$.

Thus, an embedding is the same thing as a strong homomorphism, which is one-to-one. A structure defined for all formulas in the language consisting of the language of A together with a constant symbol for each element of M, which is interpreted as that element. A structure is said to be a **model** of a theory T if the language of M is the same as the language of T and every sentence in T is satisfied by M. Thus, for example, a "ring" is a structure for the language of rings that satisfies each of the ring axioms, and a model of ZFC set theory is a structure in the language of set theory that satisfies each of the ZFC axioms. Two structures M and N of the same signature σ are **elementarily equivalent** if every first-order sentence (formula without free variables) over σ is true in M if and only if it is true in N, that is, if M and N have the same complete first-order theory. If M and N are elementarily equivalent, written $M \equiv N$.

A first-order theory is complete if and only if any two of its models are elementarily equivalent. N is an elementary substructure of M if N and M are structures of the same signature Σ such that for all first-order Σ-formulas $\varphi(x_1, \ldots, x_n)$ with free variables $x_1, \ldots, x_n$, and all elements $a_1, \ldots, a_n$ of N, $\varphi(a_1, \ldots, a_n)$ holds in N if and only if it holds in M: $N \models \varphi(a_1, \ldots, a_n)$ iff $M \models \varphi(a_1, \ldots, a_n)$. Let M be a structure of signature R(a1, …, an), there are $b_1, \ldots, b_n$ such that $R(h(a_1), \ldots, h(a_n))$ and N a substructure of M. N is an elementary substructure of M if and only if for every first-order formula $\varphi(x, y_1, \ldots, y_n)$ over σ and all elements $b_1, \ldots, b_n$ from N, if M $x\backslash\varphi(x, b_1, \ldots,$

b_n), then there is an element a in N such that M $\varphi(a, b_1, \ldots, b_n)$. An *elementary embedding* of a structure *N* into a structure *M* of the same signature is a map h : *N* → *M* such that for every first-order σ-formula $\varphi(x_1, \ldots, x_n)$ and all elements $a_1, \ldots, a_n$ of *N*, *N* |= $\varphi(a_1, \ldots, a_n)$ implies M |= $\varphi(h(a_1), \ldots, h(a_n))$. Every elementary embedding is a strong homomorphism, and its image is an elementary substructure.

Proposition 2.2 Let A and B be models for a language L. Then A is isomorphically embedded in B iff B can be expanded to a model of the diagram of A.

As observed in the section on first-order logic, first-order theories cannot be categorical, that is, they cannot describe a unique model up to isomorphism, unless that model is finite. But two famous model-theoretic theorems deal with the weaker notion of κ-categoricity for a cardinal κ. A theory T is called **κ-categorical** if any two models of T that are of cardinality κ are isomorphic. It turns out that the question of κ-categoricity depends critically on whether κ is bigger than the cardinality of the language (i.e., $\aleph_0 + |\sigma|$, where $|\sigma|$ is the cardinality of the signature). For finite or countable signatures this means that there is a fundamental difference between $\aleph_0$ -cardinality and κ-cardinality for uncountable κ.

A few characterizations of **$\aleph_0$-categoricity** include:

For a complete first-order theory T in a finite or countable signature the following conditions are equivalent:

1. *T* is $\aleph_0$ -categorical.
2. For every natural number *n*, the Stone space *Sn*(*T*) is finite.
3. For every natural number *n*, the number of formulas $\varphi(x_1, \ldots, x_n)$ in n free variables, up to equivalence modulo T, is finite.

This result, due independently to Engeler, Ryll-Nardzewski and Svenonius, is sometimes referred to as the Ryll-Nardzewski theorem.

Further, ℵ0 -categorical theories and their countable models have strong ties with oligomorphic groups. They are often constructed as Fraïssé limits.

Michael Morley's highly nontrivial result that (for countable languages) there is only one notion of uncountable categoricity was the starting point for modern model theory, and in particular classification theory and stability theory:

Morley's categoricity theorem: If a first-order theory T in a finite or countable signature is κ-categorical for some uncountable cardinal κ, then T is κ-categorical for all uncountable cardinals κ (c.f. Chapters 4).

Uncountably categorical (i.e., κ-categorical for all uncountable cardinals κ) theories are from many points of view the most well behaved theories. A theory that is both $\aleph_{0}$-categorical and uncountable categorical is called totally categorical.

2.7 PRODUCTS ON MODELS

The ultraproduct construction is a uniform method of building models of first order theories, which has applications in many areas of mathematics. It is attractive because it is algebraic in nature, but preserves all properties expressible in first order logic. The idea goes back to the construction of nonstandard models of arithmetic by Skolem [51] in 1934. In 1948, Hewitt studied ultraproducts of fields. For first order structures in general, the ultraproduct construction was defined by (Lo's, 1955). The subject developed rapidly beginning in 1958 with a series of abstracts by Frayne, Morel, Scott, and Tarski, which led to the 1962 paper, other early papers are by Kochen, and Kiesler. The groundwork for the application of ultraproducts to mathematics was laid in the late 1950s through the 1960s. The purpose of this article is to give a survey of the classical results on ultraproducts of first order structures in order to provide some background for the papers in this volume. Over the years, many generalizations of the ultraproduct construction, as well as applications of ultraproducts to nonfirst order structures, have appeared in the literature. To keep this paper of reasonable length, we will not include such generalizations in this survey. For earlier surveys of ultraproducts see Kiseler (1967).

We assume familiarity with a few basic concepts from model theory. The cardinality of a set X is denoted by $|X|$. The cardinality of N is denoted by ω. The set of all subsets of a set I is denoted by P(I), and the set of finite subsets of I by Pω(I). Given mappings $f : X \rightarrow Y$ and $g : Y \rightarrow Z$, the composition $g \circ f : X \rightarrow Z$ is the mapping $x \rightarrow g(f(x))$. A first order vocabulary L consists of a set of finitary relation symbols, function symbols, and constant symbols. We use A, B,... to denote L-structures with universe sets A, B,.... By the cardinality of A we mean the cardinality of its universe set A.

The notation $A |= \varphi(a_1,..., a_n)$ means that the formula $\varphi(x_1,..., x_n)$ is true in A when each xi is interpreted by the corresponding ai. The notation h : A→B means that h is a homomorphism of A into B, that is, h maps A into B and each atomic formula which is true for a tuple in A is true for the h-image of the tuple in B. The notation h : A $\subseteq$ B means that h is an (isomorphic) embedding of A into B, that is, h maps A into B and each quantifier-free formula of L which is true for a tuple in A is true for the h-image of the tuple in B. h : A~=B means that h is an isomorphism from A onto B, and A ~= B means that A and B are isomorphic.

The set of all sentences true in A is called the complete theory of A. A and B are called elementarily equivalent, in symbols A ≡ B, if they have the same complete theory. The notation h : A ≺ B means that h is an elementary embedding from A into B, that is, h maps A into B and each formula of L which is true for a tuple in A is true for the h-image of the tuple in B. Clearly, h : A≺B implies that A ≡ B. We say that B is an elementary extension of A and write A ≺ B if A ≺ B and the identity map is an elementary embedding of A into B. It is easy to see that if h : A □ B, then B is isomorphic to some elementary extension of A.

A fundamental result that is used very often in model theory is the compactness theorem, which says that if every finite subset of a set T of sentences has a model, then T has a model. One application of compactness is the construction of extremely rich models called saturated models. An L-structure A is said to be κ— saturated if every set of first order formulas with fewer than κ parameters from A which is finitely satisfied in A is satisfied in A. A is saturated if it is |A|-saturated. Morley and Vaught (Chapter 8) proved that any two elementarily equivalent saturated structures of the same cardinality are isomorphic, that each infinite structure A has a saturated elementary extension in each inaccessible cardinal $\kappa \geq |A| + |L|$, and has a κ+-saturated elementary extension of cardinality 2^κ whenever $2^\kappa \geq |A|$ and $\kappa \geq |L|$.

Given two vocabularies L1 $\subseteq$ L2, the reduct of an L2-structure A2 to L1 is the L1-structure A1 obtained by forgetting the interpretation of each symbol of L2/L1. An expansion of an L1-structure A1 to L2 is an L2-structure formed by adding interpretations of the symbols of L2 \ L1, that is, an L2-structure whose reduct to L1 is A1. A **saturated model** M is one, which realizes as many complete types as may be "reasonably expected" given its size. For example, an ultrapower model of the hyperreals is $\aleph_1$-saturated, which means that every descending nested sequence of internal sets has a nonempty intersection Goldblatt (1998).

2.8 MODEL THEORY AND TOPOI

2.8.1 MONOIDAL CATEGORIES

Recall that a one-object category is a monoid. Now we can observe that any category is a generalized monoid. Given K, let $M_K = U_{\text{A, B in Object K}}$ K (A, B) be the collection of all K-morphisms. Composition defines a partial function (*) $M_K \times M_K \rightarrow M_{K:}$ (f, g):→ g.f. We can recapture the identities of K from this function. f has exactly one identity u, such that, u.f is defined. Let us call that C (f) for the identi of codomains of f. There is exactly one v such that f Ÿ ½ is defined. Call it D(f) for the domain of f. Thus the partial function (*) satisfies:

1. There are total functions C, D: $M_K \rightarrow$ identities of M_K for which D(D(f)) = D(f) = C(D(f)) = C(f) = (C(C(f)).
2. g. f is defined iff C(f) = D(g). If g. f is defined then D(g.f) = D(f). and C(g.f) = C(g). Moreover, if either (h.g).f or h. (g.f) is defined, then both are defined and are equal.

Define generalized homorphisms for generalized monoids

$$H = M_K \rightarrow M_L$$

Obvious equivalent is f an identity g. f is defined è Hf an identity

1. g. f is defined H(g.f) = Hg.Hf

That is, H(idA) =id $_{\text{since f.}}$ idA is defined for f : →B we have from (2) above that Hf = H(f idA) = Hf.H(idA)= HF. Id $_{\text{AH}}$

That implies HF must have domain AH and co-domain BH.

2. Says H (A –F→ B) = HA —HF→ HB.

For each object A of K denote by HA the object is which H(idA) is the identity,

Thus we can define: A functor H from a category K to a category L is a function which maps Obj (K) → Obj (L): A a HA and which for each pair

A, B of objects of K maps K(A, B) → L(HA, HB): f) → Hf while satisfying H (id_A) = id_{HA} all A in Obj(K).

H(g. f) = Hg. Hf whenever g.f is defined in K. We say that H is an isomorphism of A a HA and each K(A, B) → L(HA, HB) are bijections.

2.8.2 HEYTING ALGEBRAS

In mathematics, a **Heyting algebra**, named after Arend Heyting, is a bounded lattice (with join and meet operations written ∨ and ∧ and with least element 0 and greatest element 1) equipped with a binary operation a→b of implication such that (a→b)∧a ≤ b, and moreover a→b is the greatest such in the sense that if c∧a ≤ b then c ≤ a→b. From a logical standpoint, A→B is by this definition the weakest proposition for which modus ponens, the inference rule A→B, A |– B, is sound. Equivalently a Heyting algebra is a residuated lattice whose monoid operation a•b is a ∧ b; yet another definition is as a posetal cartesian closed category with all finite sums. Like Boolean algebras, Heyting algebras form a variety axiomatizable with finitely many equations.

As lattices, Heyting algebras can be shown to be distributive. Every Boolean algebra is a Heyting algebra when a→b is defined as usual as ¬a ∨b, as is every complete distributive lattice[clarification needed] when a → b is taken to be the supremum of the set of all c for which a∧c ≤ b. The open sets of a topological space form a complete distributive lattice and hence a Heyting algebra. In the finite case every nonempty distributive lattice, in particular every nonempty finite chain, is automatically bounded and complete and hence a Heyting algebra.

It follows from the definition that 1 ≤ 0→a, corresponding to the intuition that any proposition a is implied by a contradiction 0. Although the negation operation ¬a is not part of the definition, it is definable as a→0. The definition implies that a∧¬a = 0, making the intuitive content of ¬a the proposition that to assume a would lead to a contradiction, from which any other proposition would then follow. It can further be shown that a ≤ ¬¬a, although the converse, ¬¬a ≤ a, is not true in general, that is, double negation does not hold in general in a Heyting algebra. Heyting algebras generalize Boolean algebras in the sense that a Heyting algebra satisfying

$a \vee \neg a = 1$ (excluded middle), equivalently $\neg\neg a = a$ (double negation), is a Boolean algebra. Those elements of a Heyting algebra of the form $\neg a$ comprise a Boolean lattice, but in general this is not a subalgebra of H. Heyting algebras serve as the algebraic models of propositional intuitionistic logic in the same way Boolean algebras model propositional classical logic. Complete Heyting algebras are a central object of study in pointless topology. The internal logic of an elementary topos is based on the Heyting algebra of subobjects of the terminal object 1 ordered by inclusion, equivalently the morphisms from 1 to the subobject classifier Ω.

Every Heyting algebra with exactly one coatom is subdirectly irreducible, whence every Heyting algebra can be made an SI by adjoining a new top. It follows that even among the finite Heyting algebras there exist infinitely many that are subdirectly irreducible, no two of which have the same equational theory. Hence no finite set of finite Heyting algebras can supply all the counterexamples to nonlaws of Heyting algebra. This is in sharp contrast to Boolean algebras, whose only SI is the two-element one, which on its own therefore suffices for all counterexamples to nonlaws of Boolean algebra, the basis for the simple truth table decision method. Nevertheless it is decidable whether an equation holds of all Heyting algebras.

Heyting algebras are less often called **pseudoBoolean algebras**, or even **Brouwer lattices.**

Definition 2.10 A Heyting algebra H is a bounded lattice such that for all a and b in H there is a greatest element x of H such that $a \wedge x \leq b$.

This element is the **relative pseudocomplement** of A with respect to b, and is denoted $a \rightarrow b$. We write 1 and 0 for the largest and the smallest element of H, respectively.

In any Heyting algebra, one defines the **pseudocomplement** $\neg A$ of any element A by setting $\neg a = (a \rightarrow 0)$. By definition, $a \wedge \neg a = 0$, and $\neg a$ is the largest element having this property. However, it is not in general true that $a \vee \neg a = 1$, thus $\neg$ is only a pseudocomplement, not a true complement, as would be the case in a Boolean algebra.

A **complete Heyting algebra** is a Heyting algebra that is a complete lattice.

A **subalgebra** of a Heyting algebra H is a subset H1 of H containing 0 and 1 and closed under the operations $\wedge \vee$ and $\rightarrow$. It follows that it is also closed under $\neg$. A subalgebra is made into a Heyting algebra by the induced operations.

An equivalent definition of Heyting algebras can be given by considering the mappings $f_a : H \rightarrow H$ defined by $f_a(x) = a \wedge x$, for some fixed in H. A bounded lattice H is a Heyting algebra if and only if every mapping f_a is the lower adjoint of a monotone Galois connection. In this case the respective upper adjoint g_a is given by $g_a(x) = a \rightarrow x$, where $\rightarrow$ is defined as above.

Yet another definition is as a residuated lattice whose monoid operation is $\wedge$. The monoid unit must then be the top element 1. Commutatively of this monoid implies that the two residuals coincide as $a \rightarrow b$.

Given a bounded lattice A with largest and smallest elements 1 and 0, and a binary operation $\rightarrow$, these together form a Heyting algebra if and only if the following hold:

This characterization of Heyting algebras makes the proof of the basic facts concerning the relationship between intuitionist propositional calculus and Heyting algebras immediate. (For these facts, see the sections "Provable identities" and "Universal constructions".) One should think of the element 1 as meaning, intuitively, "provably true." With enough rules available at Heyting algebras one prove Heyting algebra assertions without applying ordinary logic deductions.

Given a set A with three binary operations $\wedge \vee$ and $\rightarrow$, with two distinguished elements 0 and 1, then A is a Heyting algebra for these operations (and the relation $\leq$ defined by the condition that $a \leq b$ when $(a \rightarrow b) = 1$ if and only if the following conditions hold for any elements x, y and z of A:

Finally, we define $\neg x$ to be $x \rightarrow 0$.

Condition 1 says that equivalent formulas should be identified. Condition 2 says that provably true formulas are closed under modus ponens. Conditions 3 and 4 are *then* conditions. Conditions 5, 6 and 7 are *and* conditions. Conditions 8, 9 and 10 are *or* conditions. Condition 11 is a *false* condition.

Of course, if a different set of axioms were chosen for logic, we could modify ours accordingly.

The free Heyting algebra over one generator (aka Rieger–Nishimura lattice)

Every Boolean algebra is a Heyting algebra, with $p \rightarrow q$ given by $\wedge$ $p \wedge q$.

Every totally ordered set that is a bounded lattice is also a Heyting algebra, where $p \to q$ is equal to q when $p > q$, and 1 otherwise.

The simplest Heyting algebra that is not already a Boolean algebra is the totally ordered set {0, ½, 1} with → defined as above.

Every interior algebra provides a Heyting algebra in the form of its lattice of open elements. Every Heyting algebra is of this form as a Heyting algebra can be completed to a Boolean algebra by taking its free Boolean extension as a bounded distributive lattice and then treating it as a generalized topology in this Boolean algebra.

The Lindenbaum algebra of propositional intuitionistic logic is a Heyting algebra.

The global elements of the subobject classifier W of an elementary topos form a Heyting algebra; it is the Heyting algebra of truth-values of the intuitionistic higher-order logic induced by the topos.

The ordering on a Heyting algebra H can be recovered from the operation → as follows: for any elements *a*, *b* of H, $a < b$ if and only if $a \to b=1$.

In contrast to some many-valued logics, Heyting algebras share the following property with Boolean algebras: if negation has a fixed point (i.e., $\neg a= a$ for some *a*), then the Heyting algebra is the trivial one-element Heyting algebra.

Given a formula F(A1, A2, …An) of propositional calculus (using, in addition to the variables, the connectives ∧∨ Ø and →., and the constants 0 and 1), it is a fact, proved early on in any study of Heyting algebras, that the following two conditions are equivalent:

The formula F is provably true in intuitionist propositional calculus.

The identity $F(a_1, a_2, \ldots, a_n) = 1$ is true for any Heyting algebra H and any elements $a_1, a_2, \ldots, a_n \in H$.

The metaimplication $1 \Rightarrow 2$ is extremely useful and is the principal practical method for proving identities in Heyting algebras. In practice, one frequently uses the deduction theorem in such proofs.

Since for any *a* and *b* in a Heyting algebra H we have $a \leq b$ if and only if $a \to b=1$, it follows from $1 \Rightarrow 2$ that whenever a formula $F \to G$ is provably true, we have $F(a_1, a_2, \ldots, a_n) =< G(a_1, a_2, \ldots, a_n)$ for any Heyting algebra H, and any elements $a_1, a_2, \ldots, a_n \in H$. (It follows from the deduction theorem that $F \to G$ is provable if and only if G is provable from F, that is, if G is a provable consequence of F. In particular, if F and G are provably equivalent, then $F(a_1, a_2, \ldots, a_n) \leq G(a_1, a_2, \ldots, a_n)$, since $\leq$ is an order relation.

$1 \Rightarrow 2$ can be proved by examining the logical axioms of the system of proof and verifying that their value is 1 in any Heyting algebra, and then verifying that the application of the rules of inference to expressions with value 1 in a Heyting algebra results in expressions with value 1. For example, let us choose the system of proof having modus ponens as its sole rule of inference, and whose axioms are the Hilbert-style ones given at Intuitionistic logic Axiomatization. Then the facts to be verified follow immediately from the axiom-like definition of Heyting algebras given above.

$1 \Rightarrow 2$ also provides a method for proving that certain propositional formulas, though tautologies in classical logic, *cannot* be proved in intuitionist propositional logic. In order to prove that some formula $F(a_1, \ldots, a_n)$ is not provable, it is enough to exhibit a Heyting algebra H and elements $a_1, a_2, \ldots, a_n \in H$ such that $F(a_1, \ldots, a_n) \neq 1$.

Heyting algebras are always distributive. Specifically, we always have the identities.

The distributive law is sometimes stated as an axiom, but in fact it follows from the existence of relative pseudocomplements. The reason is that, being the lower adjoint of a Galois connection $\wedge$ preserves all existing suprema. Distributive in turn is just the preservation of binary suprema by $\wedge$.

By a similar argument, the following infinite distributive law holds in any complete Heyting algebra:

Definition 2.11 (Heyting algebra morphisms)

Given two Heyting algebras H_1 and H_2 and a mapping $f : H_1 \rightarrow H_2$, we say that f is a **morphism** of Heyting algebras if, for any elements x and y in H_1, we have:

We put condition 6 in brackets because it follows from the others, as $\neg x$ is just $x \rightarrow 0$, and one may or may not wish to consider $\neg$ to be a basic operation.

It follows from Conditions 3 and 5 (or 1 alone, or 2 alone) that f is an increasing function, that is, that $f(x) \leq f(y)$ whenever $x \leq y$.

Assume H_1 and H_2 are structures with operations $\rightarrow$, $\wedge$, $\vee$ (and possibly $\neg$) and constants 0 and 1, and f is a subjective mapping from H_1 to H_2

with properties 1 through 5 (or 1 through 6) above. Then if H_1 is a Heyting algebra, so too is H_2. This follows from the characterization of Heyting algebras as bounded lattices (thought of as algebraic structures rather than partially ordered sets) with an operation $\rightarrow$ satisfying certain identities.

The identity map $f(x) = x$ from any Heyting algebra to itself is a morphism, and the composite $g \circ f$ of any two morphisms f and g is a morphism. Hence Heyting algebras form a category.

Given a Heyting algebra H and any subalgebra H_1, the inclusion mapping $i : H_1 \rightarrow H$ is a morphism.

For any Heyting algebra H, the map $x \mapsto \neg\neg x$ defines a morphism from H onto the Boolean algebra of its regular elements H_{reg}. This is *not* in general a morphism from H to itself, since the join operation of H_{reg} may be different from that of H.

Let H be a Heyting algebra, and let $F \subseteq H$. We call F a **filter** on H if it satisfies the following properties: If $x \in$ F, $y \in$ H, and $x \leq y$ then $y \in$ F.

The intersection of any set of filters on H is again a filter. Therefore, given any subset S of H there is a smallest filter containing S. We call it the filter **generated** by S. If S is empty, $F = \{1\}$. Otherwise, F is equal to the set of x in H such that there exist $y_1, y_2, \ldots, y_n \in S$ with $y_1 \wedge y_2 \wedge \ldots \wedge y_n \leq x$.

If H is a Heyting algebra and F is a filter on H, we define a relation $\sim$ on H as follows: we write $x \sim y$ whenever $x \rightarrow y$ and $y \rightarrow x$ both belong to F. Then $\sim$ is an equivalence relation; we write H/F for the quotient set. There is a unique Heyting algebra structure on H/F such that the canonical surjection $pF : H \rightarrow H/F$ becomes a Heyting algebra morphism. We call the Heyting algebra H/F the **quotient** of H by F.

Let S be a subset of a Heyting algebra H and let F be the filter generated by S. Then H/F satisfies the following universal property:

Given any morphism of Heyting algebras $f : H \rightarrow H'$ satisfying $f(y) = 1$ for every $y \in S$, f factors uniquely through the canonical surjection $pF : H \rightarrow H/F$. That is, there is a unique morphism $f' : H/F \rightarrow H'$ satisfying $f'pF = f$. The morphism f' is said to be *induced* by f.

Let $f : H_1 \rightarrow H_2$ be a morphism of Heyting algebras. The **kernel** of f, written ker f, is the set $f^{-1}[\{1\}]$. It is a filter on H_1. (Care should be taken because this definition, if applied to a morphism of Boolean algebras, is dual to what would be called the kernel of the morphism viewed as a morphism of rings.) By the foregoing, f induces a morphism f': $H_1/(\ker f) \to H_2$. It is an isomorphism of $H_1/(\ker f)$ onto the subalgebra $f[H_1]$ of H_2.

2.9 MORE ON UNIVERSAL CONSTRUCTIONS

Heyting algebra of propositional formulas in n variables up to intuitionist equivalence is an example to consider.

The metaimplication 2 ⇒ 1 in the section "Provable identities" is proved by showing that the result of the following construction is itself a Heyting algebra:

Consider the set L of propositional formulas in the variables $A_1, A_2, \ldots, A_n$.

Endow L with a preorder $\preccurlyeq$ by defining $F \preccurlyeq G$ if G is an (intuitionist) logical consequence of F, that is, if G is provable from F. It is immediate that $\preccurlyeq$ is a preorder.

Consider the equivalence relation $F \sim G$ induced by the preorder $F \preccurlyeq G$. (It is defined by $F \sim G$ if and only if $F \preccurlyeq G$ and $G \preccurlyeq F$. In fact, $\sim$ is the relation of intuitionist logical equivalence.

Let H_0 be the quotient set $L/\sim$. This will be the desired Heyting algebra.

We write $[F]$ for the equivalence class of a formula F. Operations $\to$, $\wedge$, $\vee$ and $\neg$ are defined in an obvious way on L. Verify that given formulas F and G, the equivalence classes $[F \to G]$, $[F \wedge G]$, $[F \vee G]$ and $[\neg F]$ depend only on $[F]$ and $[G]$. This defines operations $\to$, $\wedge$, $\vee$ and $\neg$ on the quotient set $H_0 = L/\sim$. Further define 1 to be the class of provably true statements, and set $0 = [\perp]$.

Verify that H_0, together with these operations, is a Heyting algebra. We do this using the axiom-like definition of Heyting algebras. H_0 satisfies conditions THEN-1 through FALSE because all formulas of the given forms are axioms of intuitionist logic. MODUS-PONENS follows from the fact that if a formula $\top \to F$ is provably true, where $\top$ is provably true, then F is provably true (by application of the rule of inference modus ponens). Finally, EQUIV results from the fact that if $F \to G$ and $G \to F$ are both provably true, then F and G are provable from each other (by application of the rule of inference modus ponens), hence $[F] = [G]$.

As always under the axiom-like definition of Heyting algebras, we define $\leq$ on H_0 by the condition that $x \leq y$ if and only if $x \rightarrow y = 1$. Since, by the deduction theorem, a formula $F \rightarrow G$ is provably true if and only if G is provable from F, it follows that $[F] \leq [G]$ if and only if $F \preccurlyeq G$. In other words, $\leq$ is the order relation on $L/\sim$ induced by the preorder $\preccurlyeq$ on L.

The preceding construction can be carried out for any set of variables $\{Ai: i \in I\}$ (possibly infinite). One obtains in this way the *free* Heyting algebra on the variables $\{Ai\}$, which we will again denote by H_0. It is free in the sense that given any Heyting algebra H given together with a family of its elements $\langle ai: i \in I \rangle$, there is a unique morphism $f: H_0 \rightarrow H$ satisfying $f([Ai]) = ai$. The uniqueness of f is not difficult to see, and its existence results essentially from the metaimplication $1 \Rightarrow 2$ of the section "Provable identities" above, in the form of its corollary that whenever F and G are provably equivalent formulas, $F(\langle ai \rangle) = G(\langle ai \rangle)$ for any family of elements $\langle ai \rangle$ in H.

Given a set of formulas T in the variables $\{Ai\}$, viewed as axioms, the same construction could have been carried out with respect to a relation $F \preccurlyeq G$ defined on L to mean that G is a provable consequence of F and the set of axioms T. Let us denote by HT the Heyting algebra so obtained. Then HT satisfies the same universal property as H_0 above, but with respect to Heyting algebras H and families of elements $\langle ai \rangle$ satisfying the property that $J(\langle ai \rangle) = 1$ for any axiom $J(\langle Ai \rangle)$ in T. (Let us note that HT, taken with the family of its elements $\langle [Ai] \rangle$, itself satisfies this property.) The existence and uniqueness of the morphism is proved the same way as for H_0, except that one must modify the metaimplication $1 \Rightarrow 2$ in "Provable identities" so that 1 reads "provably true *from T*," and 2 reads "any elements $a_1, a_2, \ldots, an$ in *H satisfying the formulas of T*."

The Heyting algebra HT that we have just defined can be viewed as a quotient of the free Heyting algebra H_0 on the same set of variables, by applying the universal property of H_0 with respect to HT, and the family of its elements $\langle [Ai] \rangle$. Every Heyting algebra is isomorphic to one of the form HT.

2.9.1 LINDENBAUM ALGEBRAS

The constructions we have just given play an entirely analogous role with respect to Heyting algebras to that of Lindenbaum algebras with respect to Boolean algebras. In fact, The Lindenbaum algebra BT in the variables $\{Ai\}$ with respect to the axioms T is just our $HT_{\cup}T_1$, where T_1 is the set of all formulas of the form $\neg\neg F \rightarrow F$, since the additional axioms of T_1 are the only ones that need to be added in order to make all classical tautologies provable.

Heyting algebras as applied to intuitionistic logic if one interprets the axioms of the intuitionistic propositional logic as terms of a Heyting algebra, then they will evaluate to the largest element, 1, in *any* Heyting algebra under any assignment of values to the formula's variables. For instance, $(P \wedge Q) \rightarrow P$ is, by definition of the pseudocomplement, the largest element x such that $P \wedge Q \wedge x \leq P$. This inequation is satisfied for any x, so the largest such x is 1.

Furthermore, the rule of modus ponens allows us to derive the formula Q from the formulas P and $P \rightarrow Q$. But in any Heyting algebra, if P has the value 1, and $P \rightarrow Q$ has the value 1, then it means that $P \wedge Q \leq 1$, and so $1 \wedge 1 \leq Q$; it can only be that Q has the value 1.

The converse can be proven as well: if a formula always has the value 1, then it is deducible from the laws of intuitionistic logic, so the *intuitionistically valid* formulas are exactly those that always have a value of 1. This is similar to the notion that *classically valid* formulas are those formulas that have a value of 1 in the two-element Boolean algebra under any possible assignment of true and false to the formula's variables — that is, they are formulas, which are tautologies in the usual truth-table sense. A Heyting algebra, from the logical standpoint, is then a generalization of the usual system of truth-values, and its largest element 1 is analogous to 'true'. The usual two-valued logic system is a special case of a Heyting algebra, and the smallest nontrivial one, in which the only elements of the algebra are 1 (true) and 0 (false).

2.9.2 CARTESIAN CLOSED CATEGORIES

In category theory, a category is **cartesian closed** if, roughly speaking, any morphism defined on a product of two objects can be naturally identified with a morphism defined on one of the factors. These categories are particularly important in mathematical logic and the theory of programming, in that their internal language is the simply typed lambda calculus. They are generalized by closed monoidal categories, whose internal language, linear type systems, are suitable for both quantum and classical computation.[1]

The category C is called **Cartesian closed** [2] if and only if it satisfies the following three properties:

- It has a terminal object.
- Any two objects X and Y of C have a product $X \times Y$ in C.
- Any two objects Y and Z of C have an exponential ZY in C.

The first two conditions can be combined to the single requirement that any finite (possibly empty) family of objects of C admit a product in C, because of the natural associativity of the categorical product and because the empty product in a category is the terminal object of that category.

The third condition is equivalent to the requirement that the functor $-\times Y$ (i.e., the functor from C to C that maps objects X to $X\times Y$ and morphisms φ to $\varphi\times \mathrm{id}Y$) has a right adjoint, usually denoted $-Y$, for all objects Y in C. For locally small categories, this can be expressed by the existence of a bijection between the hom-sets Hom(X x Y, Z) $\cong$ Hom (X, Z^{Y}) which is natural in both X and Z.

If a category is such that all its slice categories are cartesian closed, then it is called *locally cartesian closed.*

Examples of cartesian closed categories include:

The category **Set** of all sets, with functions as morphisms, is cartesian closed. The product $X\times Y$ is the cartesian product of X and Y, and ZY is the set of all functions from Y to Z. The adjointness is expressed by the following fact: the function $f : X\times Y \rightarrow Z$ is naturally identified with the curried function $g : X \rightarrow ZY$ defined by $g(x)(y) = f(x, y)$ for all x in X and y in Y.

The category of finite sets, with functions as morphisms, is cartesian closed for the same reason.

If G is a group, then the category of all G-sets is cartesian closed. If Y and Z are two G-sets, then ZY is the set of all functions from Y to Z with G action defined by $(g.F)(y) = g.(\mathrm{F}(g\text{-}1.y))$ for all g in G, $F{:}Y \rightarrow Z$ and y in Y.

The category of finite G-sets is also cartesian closed.

The category **Cat** of all small categories (with functors as morphisms) is cartesian closed; the exponential CD is given by the functor category consisting of all functors from D to C, with natural transformations as morphisms.

If C is a small category, then the functor category **Set** C consisting of all covariant functors from C into the category of sets, with natural transformations as morphisms, is cartesian closed. If F and G are two functors from C to **Set**, then the exponential FG is the functor whose value on the object X of C is given by the set of all natural transformations from $(X, -) \times G$ to F.

The earlier example of G-sets can be seen as a special case of functor categories: every group can be considered as a one-object category, and G-sets are nothing but functors from this category to **Set**.

The category of all directed graphs is cartesian closed; this is a functor category as explained under functor category.

In algebraic topology, cartesian closed categories are particularly easy to work with. Neither the category of topological spaces with continuous maps nor the category of smooth manifolds with smooth maps is cartesian closed. Substitute categories have therefore been considered: the category of compactly generated Hausdorff spaces is cartesian closed, as is the category of Frölicher spaces.

In order theory, complete partial orders (*cpo*s) have a natural topology, the Scott topology, whose continuous maps do form a cartesian closed category (i.e., the objects are the cpos, and the morphisms are the Scott continuous maps). Both currying and *apply* are continuous functions in the Scott topology, and currying, together with apply, provide the adjoint.[3]

A Heyting algebra is a Cartesian closed (bounded) lattice. An important example arises from topological spaces. If X is a topological space, then the open sets in X form the objects of a category O(X) for which there is a unique morphism from U to V if U is a subset of V and no morphism otherwise. This poset is a cartesian closed category: the "product" of U and V is the intersection of U and V and the exponential UV is the interior of $U \cup (X \backslash V)$.

The following categories are *not* cartesian closed:

The category of all vector spaces over some fixed field is not cartesian closed; neither is the category of all finite-dimensional vector spaces. While they have products (called direct sums), the product functors do not have right adjoints. They are, however, symmetric monoidal closed categories: the set of linear transformations between two vector spaces forms another vector space, so they are closed, and if one replaces the product by the tensor product, a similar isomorphism exists between the Hom spaces. The category of abelian groups is not cartesian closed, for the same reason.

In cartesian closed categories, a "function of two variables" (a morphism $f: X \times Y \to Z$) can always be represented as a "function of one variable" (the morphism $\lambda f: X \to ZY$). In computer science applications, this is known as currying; it has led to the realization that simply typed lambda calculus can be interpreted in any cartesian closed category.

In every cartesian closed category (using exponential notation), $(XY)Z$ and $(XZ)Y$ are isomorphic for all objects X, Y and Z. We write this as the equation,

$$(xy)z = (xz)y.$$

One may ask what other such equations are valid in all cartesian closed categories. It turns out that all of them follow logically from the following axioms:

- $x \times (y \times z) = (x \times y) \times z$
- $x \times y = y \times x$
- $x \times 1 = x$ (here 1 denotes the terminal object of C)
- $1x = 1$
- $x1 = x$
- $(x \times y)z = xz \times yz$
- $(xy)z = x(y \times z)$

Bicartesian closed categories extend cartesian closed categories with binary coproducts and an initial object, with products distributing over coproducts. Their equational theory is extended with the following axioms:

- $x + y = y + x$
- $(x + y) + z = x + (y + z)$
- $x(y + z) = xy + xz$
- $x(y + z) = xyxz$
- $0 + x = x$
- $x \times 0 = 0$
- $x^{\circ} = 1$

2.10 EXERCISES

1. Let an epi-morphism f be defined as $A \to f\!- \to B \to C$ $g.f = h.f$ $g = h$; and
⟶ a mono-morphism C —g—⟶ B —f—⟶ A $f.g = f.l$ $g.h$ —h—⟶ prove that an epi-mono factorizations are unique up to isomorphism.

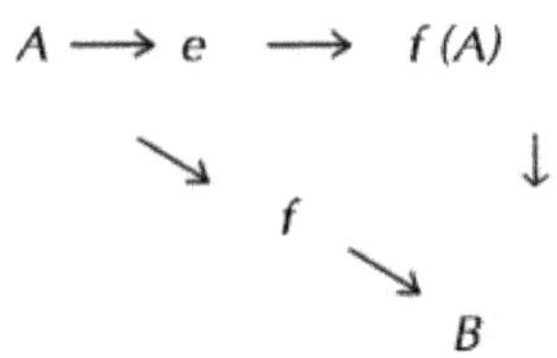

2. (i) (P_1, P_2) is a product of A_1 and A_2 iff A =~ A_1 ′ A_2 by an isomorphism transforming Pi into the map (a_1, a_2))–à a_i.
(ii) Product of two sets is unique up to isomorphism.
3. If f : A –à B then the map k: A/ ==f –à f(A): [a] —= }—à f(a) is a bijection.
4. Every coequalizer is an isomorphism.
5. Every onto map is a coequalizer.
6. Apply Yoneda's lemma to prove that Representations of functors are unique up to a unique isomorphism. That is, if (A_1, Φ_1) and (A_2, Φ_2) represent the same functor, then there exists a unique isomorphism $\varphi : A_1 \to A_2$ such that

$$\Phi_{1-}^{\ 1} \circ \Phi_2 = (j_{,} \tfrac{3}{4})$$

as natural isomorphisms from $\mathrm{Hom}(A_2, -)$ to $\mathrm{Hom}(A_1, -)$.
7. From Section 2.9.1, it is understood that that if a formula is deducible from the laws of intuitionistic logic, being derived from its axioms by way of the rule of modus ponens, then it will always have the value 1 in all Heyting algebras under any assignment of values to the formula's variables. However, one can construct a Heyting

algebra in which the value of Peirce's law is not always 1. Consider the 3-element algebra {0, ½,1} as given above. If we assign ½ to P and 0 to Q, then the value of Peirce's law $((P \rightarrow Q) \rightarrow P) \rightarrow P$ is ½. It follows that Peirce's law cannot be intuitionistically derived.

KEYWORDS

- **automorphism**
- **bimorphism**
- **endomorphism**
- **epimorphism**
- **hom-class**
- **isomorphism**
- **monomorphism**
- **morphisms**
- **retraction**
- **section**

CHAPTER 3

INFINITE LANGUAGE CATEGORIES

CONTENTS

Infinite language categories were presented and their categorical properties are defined since the authors 1994–1995 briefs. Infinite language categories and Functorial model theory are a specific alternative to a categorical interpretation for its logic in the categorical logic sense. We have started to define the computing applications for the Functorial model theory. Commencing with the infinite language category $L_{\omega 1, K}$ we define generic sets on $L_{\omega 1, K}$ and generic computable Functors. The functor from Lω1, K to Set is an example generic functor. The functor is proved to have a limit defining positive generic models. Functorial model theory and a three category play with $L_{\omega 1, K}$, Set, and D<A, G>. D<A, G> is the category for models definable on generic diagrams. The Functors define generic sets from language strings to form limits and models. Generic functors and the new techniques are applied to define functorial models on Initial ordered structures.

3.1 BASICS

This chapter defines functorial models on ordered structures. Infinite language categories and their preliminary categorical properties are presented. The techniques we have put forth are not the same as a categorical interpretation for logic as in Categorical Logic. What had not been known before "our functorial" can be replaced by "the present Functorial", is Lω1, ω's subtle categorical structure and the structure's significance for computing theory. The categorical structure defined by INFLCS [6] and its Functorial model theory is the foundations for the project. Computable Functors are defined in a December 1996 chapter [11]. There is a basis for a functorial language-structure models. The project allows us specify computation with Lω1, ω as a language for which there are prespecified functorial semantics. There are many chapters on Lω1, ω applications to computing theory. The categorical structure, however, is new and has no obvious connections to the 1980's applications. The new functorial techniques are applicable to categorical logic where while logic can be carried out at categories, defining models presents us with challenging new mathematics areas. To define solution sets for categories we had developed new techniques, which are in part categorical topology [5]. We start with a preliminary form to the adjoint functor theorem. The solution set conditions are necessary for presenting initial objects and form the basis for the above preliminary theorem and to what is referred to as the adjoint functor theorem and the basis for many fundamental results in category theory. We had presented techniques [4] for obtaining solution sets by defining a model theory with a countable fragment of an infinitary logic for categories. Further directions for research are presented as a consequence.

Start with a well-behaved countable fragment of an infinitary language Lω1, ω defined by H. J. Keisler. The term fragment is not inconsistent with the terminology applied in categorical logic. A subclass F of class of all formulas of an Infinitary language is called a fragment, if (a) for each formula φ in F all the subformulas of φ also belong to F; and (b) F is closed under substitution: if φ is in F, *t* is a term of L, *x* is a free variable in φ, the φ(*x*/*t*) is in F.

Definition 3.1 By a fragment of Lω1,ω we mean a set L<A> of formulas such that (1) Every formula of L belongs to L<A> (2) L<A> is closed under ¬, ∃ x, and finite disjunction (3) if f (x) ∈ L<A> and t is a term then f (t) ∈ L<A> (4) If f ∈ L<A> then every subformula of f ∈ L<A>

Subformulas are defined by recursion from the infinite disjunction by sub (VΦ) = ∪ sub (ϕ) ∪ {VΦ taken over the set of formulas in with the basis defined for atomic formulas} by sub(ϕ) = {ϕ}; and for compound formulas by taking the union of the subformulas quantified or logically connected with the subformula relation applied to the original formula as a singleton. Let C be a countable set of new symbols and form the first order language K by adding to L the constants c in C. We shall let K<A> be the set of all formulas obtained from formulas in L<A> by substituting finitley many free variables by constants c in C. The K<A> is the least fragment which contains L<A>. Each formulas in K<A> contains only faintly many c in C. We refer to the fragment by $L_{\omega 1,K}$.

Let us define a small-complete category $\mathbf{L}_{\omega 1,\mathbf{K}}$ from $L_{\omega 1,K}$. The category is the preorder category defined by the formula ordering defining the language fragment. The objects are small set fragments. There are three categories at play- the category $L_{\omega 1,K}$, the category **Set,** and the category **D<A, G>.** The **D<A, G>** category is the category for models definable with D<A, G> and their morphisms. It is applied towards the solution set theory presented in the chapters that follow since Fryed's adjoint factor theorem, but not in the present chapter. The techniques we are presenting by the three categories save us from having to yet develop a categorical interpretation for $\mathbf{L}_{\omega 1, \mathbf{K},}$ in categorical logic as in Lawvere's (1967) categorical formulation of logic or the current practice.

3.2 LIMITS AND INFINITARY LANGUAGES

Start from the Op category formed by $\mathbf{L}_{\omega 1, \mathbf{K}}$. Define a functor to utter the Logische Syntax der Sprache from the language category to the category **Set** until the model appears. A limit has to be created for the corresponding model to be defined from the language syntax. The following lemma is applied towards defining generic Functors for infinite language categories. We define a small-complete category $L_{\omega 1, K}$ from $L_{\omega 1, K}$. The category is the preorder category defined by the formula ordering defining the language fragment. Let us start with the basic definition:

Definition 3.2 A category is small if the set of its objects and its arrow set are small sets. Small sets are defined to be members S to a universe U, which has closure properties such that all standard set theoretic operations on S does not leave U. All ordinary mathematics is carried out at U, and known sets such as are in U. The model theory is all carried out with [1, 3] hence, there are limiting generic sets and small index sets.

Lemma 3.1 $\mathbf{L}_{\omega 1, \mathbf{K}}$ is a small-complete category.

Proof The proof applies the foundations for the model theory. To prove $\mathbf{L}_{\omega 1, \mathbf{K}}$ is small complete, that is, every small diagram has a limit, first we define a specific model-theory and toposes for the limits by defining a functor H: ω^{op} $L_{\omega 1, K}$ and create a limit in $\mathbf{L}_{\omega 1, \mathbf{K}}$ as follows. The functor H is a list of sets H_n, consisting of the strings in $\mathbf{L}_{\omega 1, \mathbf{K}}$, and functions $f_i : H_i + 1 \rightarrow H_i$. Form the product set $P_i H_i$. Let $h = \{h_0, h_1, h_2, \ldots.\}$, with each $h_n \in H_n$, and the projections

$p_n: P_i H_i$ H_n, forming the following diagram.

H0 ← H1← Hn←Hn+1←

↑↑

ΠiHi← ← include ←S= Lim H

To form a limit at a vertex for commuting with the projections take the subset S of the strings in $\mathbf{L}_{\omega 1,\mathbf{K}}$, which match under f, that is, f $s_n+1 = s_n$, for all *n*. The above A-diagram is well-defined because sets which are generic with respect to a base set, for example the intended limit S are definable, and we only have to present limits for small A-diagrams.

The completeness part of the proof, i.e. that every functor from a small category to $\mathbf{L}_{\omega 1,\mathbf{K}}$ has a limit, is only in part similar to the completeness proof for **Set.** The crux of the proof is there being limiting cones for arbitrary functors F : J → $\mathbf{L}_{\omega 1,\mathbf{K}}$ for the K<A> fragment. We can define cones to base F from {}. The natural transformations are on functors sending arbitrary objects to sets on fragment string sets, arrowed by preorder functions. Since sets that are generic with respect to the base set are always definable, the limiting cones can always be defined. Cones to base F from {} pick fragment sets. A limiting cone can always be defined with natural transformations on the preorder functors on $\mathbf{L}_{\omega 1,\mathbf{K}}$.

A functor **F:** $\mathbf{L}^{Op}_{\omega 1,\mathbf{K}}$ → **Set** is defined by a list of sets Mn and functions fn.

What might distinguish Functorial model theory is that we do not have to develop a categorical interpretation for $\mathbf{L}_{\omega 1,\mathbf{K}}$ in categorical logic for the present chapter. Thus instead of having a direct interpretation for the Infinitary logic we define an embedding from a model definable in **D<A, G>. D<A, G>** is the category of models definable with generalized diagram D<A, G> and their morphisms presented by this author in [4]. An embedding is defined by D<A, G> to the limit created from the functor defined on the Infinitary language category **Lω1, K** to **Set.** D<A, G> is similar to the diagrams defining generic models for set theory. The sets affected are the set of formulas defined by the definition of the fragment $L_{\omega 1, K}$.

3.3 GENERIC FUNCTORS AND LANGUAGE STRING MODELS

Creating limits amounts to defining generic sets on $L_{\omega 1, \omega}$ for $\mathbf{L}_{\omega 1, K}$. The present theory creates limits for Infinitary languages to define generic models with language categories. Generic sets for infinitary languages have been applied by this author to define Positive Forcing [3] and applied to categories and toposes in [5,6].

Definition 3.3 Let M be a structure for a language L, call a subset X of M a generating set for M if no proper substructure of M contains X, i.e. if M is the closure of X U {*c*(M): *c* is a constant symbol of L}. An assignment of constants to M is a pair <A, G>, where A is an infinite set of constant symbols in L and G: A → M such that {G(a): a in A} is a set of generators for M. Interpreting a by g(a), every element of M is denoted by at least one closed term of L(A). For a fixed assignment <A, G> of constants to M, the diagram of M, D<A, G>(M) is the set of basic (atomic and negated atomic) sentences of L(A) true in M. (Note that L(A) is L enriched with set A of constant symbols.)

Definition 3.4 A *generic diagram* for a structure M is a diagram D<A, G>, such that the G in definition above has a proper definition by a specific function set.

For example, Σ1 -Skolem functions are a specific function set. We abbreviate generic diagrams by G-diagrams.

Op define a functor F: $\mathbf{L}_{\omega 1, K}$ → **Set** by a list of sets M_n and functions f_n. The functor F is a list of sets Fn, consisting of (a) the sets corresponding to an initial structure on $L_{\omega 1, K}$, for example the free syntax tree structure, where to $f(t_1, t_2, .. t_n)$ in $L_{\omega 1, K}$ there corresponds the equality relation $f(t_1, \ldots, t_n)=ft_1 \ldots t_n$ in Set; (b) the functions f_i:F_i+1 → F_i.

Let us refer to the above functor by the **generic model functor** since it defines generic sets from language strings to form limits and models. The model theoretic properties are not defined in the present chapter and are presented elsewhere [12, 14] by this author. The following theorem is from Nourani [5, 8, 18].

Theorem 3.1 The generic model functor has a limit.

Proof Let us create a limit for the functor F : $\mathbf{L}_{\omega 1, \mathbf{K}} \rightarrow$ **Set** in **Set** as follows. The functor F is a list of sets F_n, consisting of (a) the sets corresponding to an initial structure on Lω1, K, for example, to $f(t_1, t_2,.. t_n)$ in Lω1, K there corresponds the equality relation $f(t_1, \ldots, t_n)=ft_1 \ldots t_n$ in Set; (b) the functions $f_i : F_i+1 \rightarrow F_i$. Form the product set $\S_i F_i$. Let $r = \{r_0, r_2, r_3, \ldots.\}$, with each $r_n \in F_n$, and its projections p_n: $\Pi \rightarrow F_n$, forming the following diagram.

FO ← F1 ← Fn ← Fn 1←

↑ ↑ ↑ ↑

ΠiFi ← ← include ← M = Lim F

To form a limit at a vertex for commuting with the projections take the subset S of the strings in S, which match under f, that is, f sn+1 = sn, for all n. The above A-diagram is well-defined because by the definitions 2.2 to 2.3 and sets which are generic with respect to a base set, for example the intended limit M are definable [1, 3].

3.4 FUNCTORIAL MORPHIC ORDERED STRUCTURE MODELS

Ordered computing structure have become a significant area for computing since Scott [16]. Diagrams have been applied to define model-theoretic computing in my chapters since [13, 24]. Applications to Functor Computability is defined in Refs. [17, 22, 23]. Practical computing with Functors had been an important area in our projects since 1978 and in Ref. [22]. ADJ categorical [10] computing and ordered algebraic definitions are based on categories, algebraic theories, and quotient structures. Functorial model computing as it has also appeared in Refs. [8, 22] to ADJ [19, 15]. ADJ [15] define an ordered Ó-algebra and their homomorphisms with an ordering where the operations are monotonic and the homomorphims which are monotonic for the ordering defined. ADJ [79] defines ordered initial [71] categories important for computing. The ordering's significance is its being operation preserving. We define a model-theoretic ordering for the initial ordered structures to reach for models for which operation preserving orderings are definable. Nourani [78] had also written for a model theory on many-sorted categorical logic and its definability with finite similarity type. The ordering we call morphic might further be computationally appealing ever since computable functors were defined by us. We define initial ordered structures with a slight change in terminology from Ref. [79], since we have not yet defined the relation between preorder algebras and admissible sets. Admissible sets are as Barwise (1968). Functorial admissible models in the admissible set sense are due to Nourani [96–97].

Definition 3.5 A preorder << on a $\sum$-algebra A is said to be morphic iff for every s in $s,s_1...s_n$ and a_i, K_i in Asi, Ksi, respectively, if ai, bi for *i* in [*n*] then, s_A (a1, …, an) << s_B (b*1*, …, K*n*).

What I have called a morphic order is called an admissible order by (Thatcher Wagner, et.al., 1979) to save confusion from ambiguity to admissible sets. The ordering's relation to admissible sets and functores are explored at (Nourani, 1997).

Proposition 3.1 If << is a morphic preorder on a $\sum$-algebra A, then the equivalence relation ~ determined by << is a congruence relation.

Proposition 3.2 The morphic quotient structure for an ordered $\sum$—algebra A, is an ordered $\sum$-algebra.

We apply function and relation definably in the following theorem. To have a glimpse as to what it implies let us see an example. Let A be a model with built-in Skolem functions. A function f ∈ A, for A the model's universe, is said to be definable iff there exits a formula j (xyz1…zn) in the language and elements a1, …, an ∈ A such that for all a, K ∈ A, A j [aba1, …, an] iff f(a)= b. We have defined generic Functors on the category $\mathbf{L}_{\omega 1, K}$ with computable hom sets, Nourani 96 [17, 14].

Definition 3.6 For functions f and g in a structure for a language L, define the morphic *preorder* f << g iff there are formulas j and y from L such that the formulas define f and g respectively, and j is a subformula of y in the sense of definition 3.1.

Define a functor F: $\mathbf{L}_{\omega 1, K}$ → **Set** by a list of sets Mn and functions fn. The functor F is a list of sets Fn, consisting of (a) the sets corresponding to a morphic preorder on $L_{\omega 1, K}$ defined by the deniability order from the subformula Preorder on $L_{\omega 1, K}$. Starting with he free syntax tree structure, where to $f(t_1, t_2,.. t_n)$ in $\mathbf{L}_{\omega 1, K}$ there corresponds the equality relation f(t1, …, tn)=ft1…tn in Set; (*b*) the functions fi:Fi+1 → Fi.

Theorem 3.2 The morphic preorder functor defines initial models for $L_{\omega 1, K}$. To prove Theorem 4.3 there are two routes. Route A applies Keisler [1], Nourani [3] and Nourani [12, 14] with new techniques similar to Robinson's consistency theorem, however, with varying languages and fragments called functorial consistency [25]. Route B is outlined on the following paragraph from Nourani [26] where functorial morphic models are defined. One consequence of Ref. [26] is a theorem, which can get us Theorem 4.3's proof. Itself is obtained by techniques, which are by no means obvious applying admissible sets and Urelements [21, 26].

Theorem 3.3 The morphic preorders on initial structures are definable by formulas, which are preserved by end extensions on fragment consistent models.

Towards the proof: Since (Nourani, 1997) European Research council brief as follows. At the Summer 1996 European logic colloquium (Nourani, 1996) had put forth descriptive computing principles, defining descriptive computable functions. Amongst the theorems at is that for A an admissible computable set, A is descriptive computable.

Generic diagrams were applied to define admissible computable sets and models.

Definition 3.7 A model is admissible iff its universe, functions, and relations are defined with or on admissible sets.

Theorem 3.4 Admissible models are obtained by taking a reduct from the admissible hull to the Skolem hull definable by a generic diagram.

ADJ (1979) defines ordered initial MacLane (1971) categories important for computing. The ordering's significance is its being operation preserving. We define a model-theoretic ordering for the initial ordered structures to reach for models for which operation preserving orderings are definable. The author's dissertation was in part on a model theory on many-sorted categorical logic and its definability with finite similarity type. That proved to be a newer look at intuitionistic forcing a year or two later and announced at ASL, Boston. The ordering we call *morphic* might further be computationally appealing ever since computable functors were defined by Nourani (1996, 1998).

Definition 3.9 For functions f and g in a structure for a language L, define the morphic *preorder* f << g iff there are formulas j and y from L such that the formulas define f and g respectively, and j is a subformula of y in the sense of definition of the fragments for infinitary logic [Keisler 73].

We can apply our generic diagrams and sets from + forcing Nourani (1982) to the S formulas (Barwsie, 1972; Kripke-Platek] by applying a theorem form (Feferman-Kreisel, 1966.) It might also put our mind at serenity with the preorder functors and initial structures on the forthcoming chapters.

Theorem 3.5 (Ferferman-Kriesel) If a formula θ is preserved under end extensions then it is equivalent to a Σ formula.

We apply the Ferferman-Kriesel theorem to check for conditions on admissible sets starting with what we call *generic rudimentary sets*. With positive forcing [Nourani 83] and fragment consistency models (Nourani, 1997), we can check if the formula belongs to a condition for fragment consistent Functorial models.

Kiesler (1971) and Barwise (1969, 1972) admissible fragments are also pertaining. We had defined functors from the language category *Lω1, K* defined on the [Keisler 73] Lω1, w fragment to the category *Set*. Its arrows are the preorder arrows and its objects the fragments. Its properties and further areas called functorial model theory, admissible models, and ordered structures are defined by Nourani [94, 97A, 97B, 97C]. We conclude with the morphic order functor theorem. The functor creates limits on the category *Set* applying the morphic preorder to *Lω1, K*.

Theorem 3.6 The morphic preorder functor defines initial models for *Lω1, K*.The model has the morphic preorder property.

Fragment consistent model techniques (Author, 1996–2005) are applied to generate Kleene models. Kleene structures have been computing language Models interest, for example, regular expression languages. The obvious structural properties are monoidal with commutative idempotent operations that are Kleene-$*$ closed (c.f. Section 3.4.3).

Theorem 3.7 Kleene structures can be granted with an intial model characterization with morphic preorders.

3.4.1 CATEGORICAL GRAMMARS

Defining a category for languages allows us to define lifts, for example, from context. The linguistics abstraction techniques the author and Lambek, for example, have developed the author has shown (Nourani, 1996), allows us to lift from context to structures for analogical reasoning and proofs with free proof tree. For example, the generic diagrams for models technique is applied at two levels for reasoning at meta-context. Models definable with G-diagrams allow free proof trees to be defined for meta-contextual reasoning (Nourani, 1997, 2007).

The diagrams further define D<A, G> categorical abstractions are defined for lifting from diagrams to categories for definable models. A third application for Categorical grammar is as close as computational linguistics has come to what we might want to refer to as Linguistics Abstraction. There are recent techniques for structurally transforming abstract syntax by applying logical rules, for example functional composition and abstraction. They are called categorical grammars (ADJ, 1975) and the (Lambek, 1959) calculus for grammars (Lambek, 1959). These techniques have formed a basis for defining natural deduction-like rules for grammars and proof techniques for abstract syntax trees.

3.4.2. FUNCTORS AND VERSATILE ABSTRACT SYNTAX-VAS

A computational linguistics with intelligent syntax and model theory is defined by Nourani (1996a; 1996b). Intelligent functions can represent agent functions, as artificial intelligence agents, or represent languages with definitions known at syntax.

A formal computable theory can be defined based on the functions defining computable models for a context and the functions carrying context around. For the VAS context foundations is it indeed possible to decrease the computational complexity of a formal system by means of introducing context. Context localizes relevant worlds, and specific computable functions define the world. Thus extraneous deductions are instant credits reducing complexity. De-contextualization is possible and might be necessary to address structural deductions. It might further be implied by paraconsistent logics (Nourani, 1999). The paraconsistent functorial syntax and model theory might be defined where, without defying classical or intuitionistic logic, we can sensibly handle inconsistent theories. Without engaging on a fundamental logical and philosophical debate, we offer paraconsistent syntax and structures on a par with a model theory. Inconsistent theories might have a syntactic lift, for which a paraconsistent model is definable. Meta-contextual reasoning, as the author has defined in 1994 in s an interesting application area. The goal is to recapture the logics in paraconsistent logics, but only to define syntax and model theories with which paraconsistency can be defined by lifts form syntax, context, and models.

3.4.3 FRAGMENT CONSISTENT KLEENE MODELS

A Kleene algebra, (Kozen 1990) is an algebra A = (A, +,0, ;, 1, ∗) such that (A, +,0) and (A, ;,1) are monoids, with + commutative and idempotent, and satisfying

(1) $a(b+c) = ab+ac$ (2) $a0=0$ (3) $1+aa{*} <= a{*}$ (4) $1+a{*}a <= a{*}$ (5) $(a+b)c = ac+bc$ (6) $ab<=b \longrightarrow a{*}b<=b0a=0$ (7) $ba<=b \longrightarrow ba{*}<=b$

Denote by KA the class of models of these axioms, and write Horn(KA) and Eq(KA) for the Horn and equational theories of KA, respectively.

Apply fragment consistent model techniques to generate Kleene models. A String ISL algebra (Nourani, 2005) is a Σ-algebra with an additional property that the signature Σ has a subsignture Ω that is only on 1–1 functions. The algebra was developed to apply to agent computing models where the 1–1 property allows the model to treat agent computing processes apart from everyday algebraic computing signatures. We can define specific ISL algebras based on specific signatures. For example, a Kleene ISL algebra is an algebra A (A, +,0,.,1, ∗) such that (A, +,0) and (A, ;,1) are monoids, with + commutative and idempotent, and satisfying the Kleene conditions.

Lemma 3.2 String ISL algebra extending a Kleene algebra A (A, +, 0,., 1, ∗) such that (A, +, 0) and (A, ;, 1) are monoids, with + commutative and idempotent, is Kleene.

Lemma 3.3 String ISL algebra homomorphically extending a algebra A (A, +, 0,., 1, ∗) such that (A, +, 0) and (A, ;, 1) are monoids, with + commutative and idempotent, is Kleene.

Theorem 3.8 Let T be a ISL language theory. T is (a) a sound logical theory iff every axiom or proof rule in T preserves the tree game degree; (b) a complete logical theory iff there is a function-set pair < F, S > defining a canonical structure M such that M has a generic diagram definable with the functions F.

From Pratt, let L = L0, L1, L2, …, Lw be the set of languages of the form Li = xj|j < i for i = 0, 1, 2, …, w; over an alphabet whose single symbol is *x*. This set is closed under the standard regular operations and has constants 0 and 1, namely L0 and L1 respectively, making it a subalgebra of the set of all languages on that alphabet and hence a model of REG. The above are the examples of computing motivations to the following proposition. A specific equational action logic Equ(ACT) (Pratt 1990), Nourani (2006) is an example application area. The preliminary statement on the applications of the preceding section to computing, for example above, is the following proposition.

Proposition 3.3 (Positive Process Fragments) We can apply fragment consistency to obtain a model for Equational product languages by defining a language category over the signature. The preorder category has language fragment sets as objects and is preordered with the preimplication ordering.

Furthermore, we can obtain product models that apply the fragment consistency techniques the reduced products on the language sequents above up to positive Horn sentences, for example, to obtain product models.

Theorem 3.9 There is a fragment consistency positive process fragment model obtained for the language sequent Gamma that is a REG Model.

Theorem 3.10 There is a fragment consistency positive process fragment model that is a Equ(ACT) Model.

3.5 EXERCISES

1. Prove Theorem 4: The morphic preorder functor defines initial models for $L_{\omega 1, K}$.
   The model has the morphic preorder property
2. Prove Position 3.2: The morphic quotient structure for an ordered Σ-algebra A, is an ordered $\tilde{\Sigma}$-algebra.
3. Following the specific on Section 3.3 outline, prove that morphic preorders on initial structures are definable by formulas, which are preserved by end extensions on fragment consistent models.
4. Comparing to Chapter 7, consider groups with signature consisting of −1, multiplication., and the identity plus constants, with group axioms.
   Given the above presentation P, the yields a generic set G. Prove that the canonical term initial algebra M is a generic model for algebraically closed group.
5. From this chapter's developments we can obtain product models that apply the fragment consistency techniques the reduced products on the language sequent above up to positive Horn sentences, for example, to obtain product models.
   Prove Theorem 3.9: There is a fragment consistency positive process fragment model obtained for the language sequent G that is a REG Model.

KEYWORDS

- **fragment**
- **functor**
- **generic model functor**
- **Infinitary language**
- **Skolem functions**

CHAPTER 4

FUNCTORIAL FRAGMENT MODEL THEORY

CONTENTS

4.1 INTRODUCTION

Starting with infinite language categories (Nourani, 1995), and Chapter 3, specific techniques for creating functorial models based on fragments are presented in this chapter.

The model bases are Fragment Consistency Models where new techniques for creating generic models are defined. Infinitary positive language categories are defined and infinitary complements to Robinson consistency from the author's preceding papers are further developed to present new positive omitting types techniques and infinitary positive fragment higher stratified computing categories. Further neoclassic model-theoretic consequences are presented in (Nourani, 2005a). Positive categories and Horn categories are new fragment categories defined and the applications to a Positive Process algebraic computing (Nourani, 2005) is briefed on algebraic topology and computing (Nourani, 2011). For example, the author defined the category $L_{P,\omega}$ to be the category with objects positive fragments and arrows the subformula preorder on formulas to present models. Start with a well-behaved countable fragment of an infinitary language $L_{\omega 1,w}$ from Chapter 3. Let $L_{\omega 1,K}$ be the least fragment of $L_{\omega 1,w}$ that contains L<A>. Each formula j in K<A> contains only finitely many *c* in C. This implies when raking leaves on the trees, there are only finite number of named branches claimed by constant names. However, the infinite trees are defined by function names. The functions define the model with the constants. From the functorial view what follows resembles to a cosmic scooping from fragments the model theoretic specifics for a functorial model theory.

4.2 GENERIC FUNCTORS AND LANGUAGE STRING MODELS

Creating limits amounts to defining generic sets on $L_{\omega 1,\omega}$ for $L_{\omega 1,K,}$ the present theory creates limits for infinitary languages to define generic models with language categories. Generic sets for infinitary languages have been applied by this author to define Positive Forcing (1982) and applied to categories and toposes since 1997. From Chapter 3, based on the Kiesler fragment language on a theory T for the above language, let T* be

T augmented with induction schemas on the generic diagram functions in the authors' papers since two decades. To have a compete glimpse on this chapter self-contained we carry on as follows

Define a functor F: $\mathbf{L}^{Op}_{\omega 1, K}$ **Set** by a list of sets Mn and functions fn. The functor F is a list of sets Fn, consisting of:

(a) the sets corresponding to an initial structure on L1, B, for example the free syntax tree structure, where to $f(t_1, t_2,.., t_n)$ in $L_{\omega 1, K}$ there corresponds the equality relation $f(t_1, \ldots, t_n)= ft_1, \ldots, t_n$ in Set;

(b) the functions fi : Fi + 1 Fi.

Let us refer to the above functor by the generic model functor since it defines generic sets from language strings to form limits and models. The model theoretic properties are not defined in alternate chapters. The following theorem is from Nourani (1996, 2005).

Theorem 4.1 The generic model fragment string functor has a limit.

Proof

Let us create a limit for the functor F: $\mathbf{L}^{Op}_{\omega 1, K}$ Set in Set as follows. The functor F is a list of sets Fn, consisting of:

(a) the sets corresponding to an initial structure on $L_{\omega 1, K}$, for example, to $f(t_1, t_2, \ldots, t_n)$ in $\mathbf{L}_{\omega 1, K}$ there corresponds the equality relation $f(t_1, \ldots, t_n) =$ ft1, …, tn in Set;

(b) the functions fi:Fi+1 Fi. Form the product set ΠiFi. Let $r = \{r_0, r_2, r_3, \ldots\}$, with each rn $\in$ Fn, and its projections pn: ΠiFi Fn, forming the following diagram.

$A_0 < A_1 < A_2 <$ ---------

\ f1 I g1 \f2 I g2

$B_0 < B_1 <$ b2 < -----

To form a limit at a vertex for commuting with the projections take the subset S of the strings in S, which match under f, that is, f sn + 1 = sn, for all n. The above A-diagram is well defined because by the definitions 2.2 to 2.3 and sets which are generic with respect to a base set, for example, the intended limit M are definable (see Keisler, 1971; Nourani 1997).

4.3 FUNCTORIAL MODELS AS ω-CHAINS

Like onto skating on a thin functorial linguistics layer, a 1990's project by the author, c.f. Chapter 3, let us define functorial string models. We start to define functorial models by infinite chains, back and forth designs and elementary diagrams. We start to define functorial models by infinite chains, back and forth designs, and elementary diagrams. A chain of models is an increasing sequence of models $A_0 \subset A_1 \subset A_3 \subset A_4 \subset \ldots A_\beta$ $\beta<\alpha$, whose length is an ordinal α. The union of chains is the model $A = \cup A\beta$, defined as follows.

Defining ω-chain models, we take the union of chains to be the model $A = \cup A_\beta$ as follows. The universe of A is the set $A = \cup A_\beta$. Each relation R of A is the union of the corresponding relations of $A_\beta R = \cup R_\beta$. Similarly, each function G of A is the union of the corresponding functions of $A\beta$, $G = \cup G_\beta$, $\alpha < \beta$. The models $A\beta$ and A all have the same constants. An elementary chain is A chain of models $A0 \subset A1 \subset A3 \subset A4 \subset \ldots A\beta$ such that $A\gamma < A\beta$ whenever $\gamma < \beta < \alpha$, $<$ is the elementary extension relation.

Elementary chains are applied to define Robinson's consistency theorem and we plan to define similar techniques for defining String Models on the infinitary $L_{\omega 1, K}$ fragments. Of course the task is quite difficult and intricate functorial models on language fragments have to be designed.

From basic model theory we have the following theorem and the chain lemma on the following section.

Theorem 4.1 (Robinson's Consistency Theorem)

Let L1, L2 be two languages. Let $L = L1 \cap L2$. Suppose T is a complete theory in L and $T1 \supset T$, $T2 \supset T$ are consistent in L1, L2, respectively. Then $T1 \cup T2$ is consistent in the language $L1 \cup L2$.

We want to define functorial models piecemeal from language fragments by an infinite limit. There are two ways to view it.

(a) Take $L_{\omega 1, K}$ language fragments, define w-chain models, and back and forth to a limit diagram model.

(b) Define models for the Fi from elementary diagrams, that is, define a limit model by embedding from a D<A, G> model.

What complete theory can we fall onto? It has to be the theory Th(F), where F is the generic model functor defined to Set, that is,

$$\text{Th}\ (\text{F}: \mathbf{L}^{\text{Op}}_{\omega 1,\mathbf{K}} \rightarrow \mathbf{Set}).$$

4.4 MODELS GLIMPSES FROM FUNCTORS

Starting with $L_{\omega 1, K}$ language fragments, define w-chain models, and back and forth to a limit model.

Remember we had defined the fragment by letting C be a countable set of new constants symbols and form the first order language K by adding to L the constants c in C. Let K<A> be the set of formulas obtained from formulas j in L<A> by replacing finitely many free variables by constants c in C.

Lemma 4.1 For a chain $A_\beta = \beta < \alpha$, of models, $\cup A_{\beta,}$ is the unique model with universe $\cup|A_\beta|$ which contains each $A_{\beta,}$ as a submodel.

A functor F: $\mathbf{L}^{\text{Op}}_{\omega 1, \mathbf{K}} \rightarrow$ Set can be defined by sets Fi, where the Fi's are defining a free structure on some subfragment of $L_{\omega 1, K}$. To be specific we can define the subfragment models A(Fi) straight from the w-inductive definition of the Infinitary fragment. F_0 assigns names to the Set members, for example. F_1 can define 1-place functions and relations, so on and so forth.

Taking the languages defined in Section 4.1 and the Robinson's consistency theorem define Functorial Limit Chain models as follows. We shall refer to it by *FLC-models* (Nourani, 1996).

Let A and B, be models for Fi and Fi+1, respectively. Let $A \subseteq_L B$, and f: $A <_L B$ mean the L-reduct of A and B are elementarily equivalent and that f is an elementary embedding of A|L into B|L. Let A FLC model be the limit model defined by the elementary chain on the L-reducts of the models defined by the Fi's. A specific FLC model is defined by Theorem 4.2s proof.

Theorem 4.2 There is an elementary chain FLC model for L, where L is $L_{\omega 1, K}$.

Proof Let T<i, i+1> be the complete theory in L<i, i+1> = Li Ç Li+1, defined by Th (A(Fi) Ç A(Fi+1)). Let Ti and Ti+1 be arbitrary consistent theories for the subfragments Li and Li+1, respectively, satisfying Ti É T<i, i+1> and Ti+1 É T<i, i+1>. Let Ai be A(Fi) and Bi be A(Fi+1). Starting with a basis model A_0, A_0|L<0,1> and B_0|L<0,1> are models of a complete theory, hence, A_0|L<0,1> ° B_0|l<0,1>. It follows that the elementary diagram of A0|l<0,1> is consistent with the elementary diagram of B_0|L<0,1>. Therefore, there are elementary extensions B_1> B_0 and an embedding f1: A0 < B1 at L<01>. Passing to the expanded language L<0,1>A_0, we have (A_0, a)ae A_0 ° L<0,1> A_0 (B_1, fa) a ∈ A0. g1 inverse is an extension of f1.

Iterating, we obtain the tower depicted sideways.
A_0 < A_1< A_2<...
\ f1 |g1 \f2 |g2
B_0 < B_1 < b2<...

Slalom between the language pairs Li Li+1 gates to a limit model for L. For each m, fm ⊏ inverse(gm) ⊏ fm+1, fm: Am-1 < Bm at L<m-1, m>. Let A = ∪ Am, m < ω, B= ∪ Bm, m <w. B is isomorphic to a model B' such that A|L = B'|L. Piecing A and B' together we obtain a model M for $L_{\omega 1, K}$.

Note that the language L varies in our proof, while it does not in the w-chain model proof for Robinson's consistency theorem. Thus the techniques are perhaps new, inspite Barwise's 1970's accomplishments on infinitary languages.

4.5 STRUCTURE PRODUCTS

4.5.1 HORN PRODUCTS

The propositions in Section 4.5.1 Propositions 4.1–4.3 are from basic model theory. Proposition 4.4 and Theorem 4.1 is from Nourani (2005).

Proposition 4.1 Let φ be a universal sentence. Then φ is a (finite) direct product sentence iff φ is equivalent to a universal Horn sentence.

Proposition 4.2 Let φ be a universal sentence. Then φ is a (finite) direct product sentence if and only if φ is equivalent to a universal Horn sentence.

Proposition 4.3 Let φ be an existential sentence. Then φ is a (finite) direct product sentence if φ is equivalent to an existential Horn sentence.

Recall from alternate chapters that a companion closure generic filter was characterized with a set T*. That is applied in some areas in this section to create product models.

Definition 4.4 A formula is said to be positive iff it is built from atomic formulas using only the connectives ∧, v and the quantifiers.

Definition 4.5 A formula j $(x_1, x_2, \ldots, x_n)$ is preserved under homomorphisms iff for any homomorphisms f of a model A onto a model B and all a1, …, an in A if A $\models$ [a1, …, an] B $\models$ $[fa_1, \ldots, fa_n]$.

Theorem 4.3 A consistent theory is preserved under homorphisms iff T has a set of positive axioms.

Definition 4.6 A formula is said to be positive iff it is built from atomic formulas using only the connectives ∧, v and the quantifiers.

Definition 4.7 A formula j $(x_1, x_2, \ldots, x_n)$ is preserved under homomorphisms iff for any homomorphisms f of a model A onto a model B and all a1, …, an in A if A $\models$ [a1, …, an] B $\models$ $[fa_1, \ldots, fa_n]$.

Theorem 4.4 A consistent theory is preserved under homorphisms iff T has a set of positive axioms.

Proposition 4.4 Let I be the set T*. Let φ $(x_1, \ldots, x_n)$ be a Horn formula and let $\mathfrak{R}i$, i ∈ I be models for language L. Let $a_1, \ldots, a_n \in \Pi i \in I\, Ai$. The $\mathfrak{R}i$ are fragment Horn models.

If {i ∈ I: $\mathfrak{R}i \models$ φ[(a1(i) …an(i)]} then the direct P_D on $\mathfrak{R}i \models$ φ[a1D … anD], where D is the generic filter on T*.

4.5.2 POSITIVE MORPHISMS AND MODELS

Definition 4.8A formula is said to be positive iff it is built from atomic formulas using only the connectives '&', v and the quantifiers.

Definition 4.9 A formula $j(x_1, x_2, \ldots, x_n)$ is preserved under homomorphisms iff for any homomorphisms f of a model A onto a model B and all a1, …, an in A if A $\models$ [a1, …, an] B $\models$ $[fa_1, \ldots, fa_n]$.

Theorem 4.5 A consistent theory is preserved under homomorphisms iff T has a set of positive axioms.

Thus, if we apply positive formulas we can define a functorial model theory with initial models, where the arrows are homomorphisms. For what we have in mind we don't really need to apply nonpositive formulas. The functorial model theory, in its most abstract form, does not have to become specific to the formulas and what sort of morphisms.

4.6 HIGHER STRATIFIED CONSISTENCY AND COMPLETENESS

From basic mathematical logic stratification is any consistent assignment of numbers to predicate symbols guaranteeing that a unique formal interpretation of a logical theory exists. Specifically, a set of clauses of the form $q_1 \wedge \ldots \wedge q_n \wedge \emptyset\, q_{n+1} \ldots \wedge q_{n+1} \rightarrow P$ stratified if and only if there is a stratification assignment S that fulfills the following conditions:

1. If a predicate P is positively derived from a predicate p (i.e., P is the head of a rule, and p occurs positively in the body of the same rule), then the stratification number of P must be greater than or equal to the stratification number of p.
2. If a predicate P is derived from a negated predicate q (i.e., P is the head of a rule, and p occurs negatively in the body of the same rule), then the stratification number of P must be greater than the stratification number of q.

Stratified negation leads to effective techniques for stratified programs in terms of the stratified least fixpoint that is obtained by iteratively applying the fixpoint, from the lowest one up. Stratification is useful for guaranteeing unique interpretation of Horn clause theories.

In set theories, a formula in the language of first-order logic with equality and membership is said to be stratified if and only if there is a function g which sends each variable appearing in f (considered as an item of syntax) to a natural number in such a way that any atomic formula $x \in y$ appearing in ϕ satisfies $g(x) + 1 = g(y)$ and any atomic formula $x=y$ appearing in ϕ satisfies $g(x) = g(y)$.

Higher stratification is defined here based on the ordering on the fragment formula where there is an implicit assignment to the fragments on a preorder. That is how model consistency is stratified as follows.

A functor F: $\mathbf{L}_{\omega 1, K}{}^{Op} \rightarrow$Set can be defined by sets Fi, where the Fi's are defining a free structure on some subfragment of $L_{\omega 1, K}$. To be specific we can define the subfragment models A(Fi) straight from the inductive definition of the Infinitary fragment.

Let *A* be an algebraic structure or equivalently a structure in the sense of model theory, organized as a set *X* together with an indexed family of operations and relations φ_i on that set, with index set *I*. Then the reduct of *A* defined by a subset *J* of *I* is the structure consisting have the set *X* and *J*-indexed family of operations and relations whose *j*-th operation or relation for $j \in J$ is the *j*-th operation or relation of *A*. That is, this reduct is the structure *A* with the omission of those operations and relations φi for which *i* is not in *J*. Structure *A* is an expansion of *B* just when *B* is a reduct of *A*. That is, reduct and expansion are mutual converses.

Taking the languages above and the Robinson's consistency theorem let us define Functorial limit chain models, FLC, as follows. Let a *FLC* model be the limit model defined by the elementary chain on the L–reducts of the models defined by the Fi's. A specific *FLC* model is defined by the following.

Theorem 4.5 (Fragment Model Chain Consistency) Let T<i, i+1> be the complete theory in L<i, i+1> = Li Li+1, defined by Th (A(Fi) A(Fi+1)). Let Ti and Ti+1 be arbitrary consistent theories for the subfragments Li and Li+1, respectively, satisfying Ti T<i, i+1> and Ti+1 T<i, i+1>. Let Ai be A(Fi) and Bi be A(Fi+1):

(i) There are iterated elementary extensions Bi+1> Bi and an embedding tower depicted sideways.

A0 < A1< A2<......... \ f1 |g1 \f2 |g2
B0 < B1 < b2<.....
fi : Ai < Bi on the

(ii) Slaloming between the language pairs <Li, Li+1> to a limit model for L, a model M for $L_{\omega 1, K}$ can be obtained.

4.7 FRAGMENT POSITIVE OMITTING TYPE ALGEBRAS

Applying the definitions in the above sections for positive formulas we can prove the following based on Chapter 3 and Nourani (1995).

Theorem 4.6 There is an elementary chain *FLC* model for L, where L is $L_{\omega 1, K}$.

Theorem 4.7 The embedding to form elementary chains on Fi's can be defined by a back and forth model design from the strings language fragments.

Theorem 4.8 By defining models corresponding to the Fi on the fragments as an −chain from the elementary diagrams on the Th (A(Fi)) a generic model is defined by the limit

Let us start from certain model-theoretic premises with Propositions 4.5 and 4.6 known form basic model theory.

Proposition 4.5 Let $\mathfrak{R}$ and |B be models for L. Then $\mathfrak{R}$ is isomorphically embedded in B iff B can be expanded to a model of the diagram of $\mathfrak{R}$.

Proposition 4.6 Let $\mathfrak{R}$ and |B be models for L. Then $\mathfrak{R}$ is homomrphically embedded in B iff B can be expanded to a model of the positive diagram of $\mathfrak{R}$.

Further specific example embedding's from computing is Ehrig (1977).

Definition 4.10 Let Σ be a set of formulas in the variables $x_1, \ldots, x_n$. Let R be a model for L. We say that R realizes Σ iff some n-tuple of elements of A satisfies Σ in R. R omits Σ iff R does not realize Σ.

Let Σ $(x_1, \ldots, x_n)$ be a set of formulas of L. A theory T in L is said to locally realize Σ iff there is a formula φ $(x_1, \ldots, x_n)$ in L s.t.

(i) φ is consistent with T;

(ii) for all $\sigma \in \Sigma$, $T \models \varphi \rightarrow \sigma$.

That is every n-tuple of T which realizes φ satisfies Σ. We say that T locally omits Σ iff T does not locally realize Σ. For our purposes we define a new reliability basis (Nourani, 2006).

Definition 4.11 Let Σ $(x_1, \ldots, x_n)$ be a set of formulas of L. Say that a positive theory T in L positively locally realize Σ iff there is a formula φ $(x_1, \ldots, x_n)$ in L s.t. (i) φ is consistent with T : (ii) for all $\sigma \in \Sigma$, $T \models \varphi$ or T $\cup\, \sigma$ is not consistent.

Definition 4.12 Given models A and B, with generic diagrams DA and D_B we say that D_A homomorphically extends D_B iff there is a homomorphic embedding f: $A \rightarrow B$.

Consider a complete theory T in L. A formula φ $(x_1, \ldots, x_n)$ is said to be complete (in T) iff for every formula $\psi(x_1, \ldots, x_n)$, exactly one of $T \models \varphi \rightarrow \psi$ or $T \models \varphi \rightarrow \neg\, \psi$. A formula θ $(x_1, \ldots, x_n)$ is said to be completable (in T) iff there is a complete formula φ $(x_1, \ldots, x_n)$ with T models $\varphi \rightarrow \theta$. If that can't be done θ is said to be incompletable.

Theorem 4.5 Let L1, L2 be two positive languages. Let $L = L1 \cap L2$. Suppose T is a complete theory in L and $T1 \supset T$, $T2 \supset T$ are consistent in L1, L2, respectively. Suppose there is model M definable from a positive diagram in the language $L1 \cup L2$ such that there are models M1 and M2 for T1 and T2 where M can be homomorphically embedded in M1 and M2. Then

(i) $T1 \cup T2$ is consistent.

(ii) There is model N for $T1 \cup T2$ definable from a positive diagram that homomorphically extends that of M1 and M2.

Theorem 4.6 Let L1, L2 be two positive languages. Let $L = L1 \cap L2$. Suppose T is a complete theory in L and $T1 \supset T$, $T2 \supset T$ are consistent in L1, L2, respectively. Then,

(i) $T1 \cup T2$ has a model M, that is a positive end extension on Models M1 and M2 for T1, and T2, respectively.

(ii) M is definable from a positive diagram in the language $L1 \cup L2$.

Recall that a companion closure generic filter was characterized with a set T*.

Lemma 4.2 Every formula on the presentation P is completable in T*.

Let $L_{P,\omega}$ be the positive fragment obtained from the Kiesler fragment.

Define the category $L_{P,\omega}$ to be the category with objects positive fragments and arrows the subformula preorder on formulas.

Op

Define a functor F: $\mathbf{L}_{P,\omega}$ → Set by a list of sets Mn and functions fn. The functor F is a list of sets Fn, consisting of:

(a) the sets corresponding to an initial structure on $L_{P,\omega}$, for example the free syntax tree structure, where to $f(t_1, t_2,\ldots t_n)$ in $L_{P,\omega}$ there corresponds the equality relation $f(t_1, \ldots, t_n) = ft_1, \ldots, t_n$ in Set;

(b) the functions fi: Fi+1 → Fi.

Proposition 4.7 (positive morphic extensions)

The following are equivalent:

(a) Every positive sentence holding on R also holds on |B.

(b) There are elementary extensions R < R,’ |B < |B’ such that |B’ is a homomorphic image of R

Theorem 4.7 (Infinitary Fragment Consistency on Algebras) Let T<i, i+1> be complete theories for L<i, i+1> = Li intersect Li+1. Let Ti and Ti+1 be arbitrary consistent positive theories for the subfragments Li and Li+1, respectively, satisfying Ti contains T<i, i+1> and Ti+1 contains T<i, i+1>. Let Ai be A(Fi) and Bi be A(Fi+1).

(i) There are iterated elementary extensions Bi+1> Bi and an embedding fi: Ai < Bi;

(ii) Slalom between the language pairs Li Li+1 gates to a limit model for L, a model M for $L_{P,\omega}$.

Proof Starting with a basis model R<0,1> and B<0,1>, models of T0 and T1. R0|L<0,1> and B0|L<0,1> are models of a complete theory, therefore R0|L<0,1> ≡ B0|L<0,1>. It follows that the elementary diagram of A0|l<0,1> is consistent with the elementary diagram of B0|L<0,1>. Let T<i, i+1> be the positive theory in L<i, i+1> = Li intersect Li+1 defined by T* (A(Fi)) intersect A(Fi+1)), where T* is the theory obtained from the positive theory T augment with inductive consequences. We shall construct a limit model M realizing (i) enroute. Every $\Sigma(x_1, \ldots, x_n)$ a set of formulas of L<0,1> that is provable in T, T*(A0|L<0,1>) positively locally realizes Σ. That is, there is a formula $\varphi(x_1, \ldots, x_n)$ in L<0,1> s.t. φ is consistent with T*(A0|L<0,1>) and for all σ Î Σ, T* ⊨ φ or T* ∪ { σ } is not consistent. Every finite subset Σ of T*(A0|L<0,1>) has a model, therefore T*(A0|L<0,1>) has a model M0, where for every φ ∈ L <0,1> T*(A0|L<0,1>) ⊢ φ iff M0 ⊨ φ. Starting with a basis models R0 and B0, we construct a positive tower. On the iteration we realize that by s preceding proposition at each stage there are elementary extensions to Ri and Bi there are elementary extensions R < R,’ |B < |B’ such that |B’ is a homomorphic image of R’. Therefore, there are elementary extensions $B_1 > B_0$ and an embedding f1: $A_0 < B_1$ at L<0,1>. Passing to the expanded language L<0,1>A_0, we have (A_0, a) a ∈ A_0 ≡ L<0,1> A_0 (B_1, fa) a ∈ A_0. g1 inverse is an extension of f1.

Iterating, we obtain the tower depicted sideways.

$A_0 < A_1 < A_2 <$ ---------

\ f1 | g1 \f2 | g2

$B_0 < B_1 <$ b2 < -----

Slalom between the language pairs Li Li+1 gates to a limit model for L. For each m, fm ⊂ inverse (gm) ⊂ fm+1, fm: Am–1 < Bm at L < m–1, m>. Let A = ∪Am, m<ω, B = ∪Bm, m < ω. B is isomorphic to a model B’ such that A|L = B’|L. Piecing A and B’ together we obtain a model M for $L_{P,\omega}$.

4.8 OMITTING TYPES AND REALIZABILITY

Omitting types is an important technique on model-theoretic forcing and set-theoretic considerations. On the preceding section we have examined positive omitting types that this author developed on the decades publications towards a basis for topos model characterizations, natural for algebraic theories. The more general techniques are presented here that has not had direct application to Topos.

Given a theory T and a nonnegative integer n, let n(T) be the set of all formulas $\varphi(x_1,..., x_n)$ with no free variables other than x1,..., xn, such that signature (φ) ⊆ signature (T). Say that ⊆ n(T) is consistent over T if there exist A s.t. A ⊨ T and a1, …, an ∈|A| such that A ⊨ φ(a1, …, an) for all φ∈Y. Note that by compactness, if each finite subset of is consistent over T then so is.

A realization of ⊆ n (T) is an n-tuple a1, …, an ∈ |A|, A s.t. A ⊨ T, and A ⊨ φ(a1, …, an) for all φ ∈ . We then say that is realized in the model A. We say that ψ ∈ n (T) is a logical consequence of Y ⊆ n (T) over T if every realization of Y is a realization of .

Definition 4.13 An n-type over T is a set p ⊆ n(T) which is consistent over T and closed under logical consequence over T.

Definition 4.14 An n-type p over T is said to be complete if, for all φ ∈ n(T) either φ ∈ p or ¬φ ∈ p. The set of all complete n-types over T is denoted $_n$(T).

Definition 4.15 A type *p*(***x***) is said to be **isolated by** j if there is a formula j (***x***) with the property that ∀y(*x*) ∈ p(*x*), j(*x*) → y(*x*). Since finite subsets of a type are always realized in M, there is always an element ***b*** ∈ *Mn* such that j (***b***) is true in M; that is, M models j, thus ***b*** realizes the entire isolated type. So isolated types will be realized in every elementary substructure or extension. So isolated types can never be omitted.

A direct statement for omitting types is as follows:

Theorem 4.8 Omitting Types Theorem: Let L be a countable language. Let T be an L-theory. Let {pi: i ∈ N } be a countable set of nonisolated n-types of T. There is a countable L-structure M such that M models T and M omits each p_i.

Definition 4.16 Let p be an n-type over a theory T. A model A of T is said to omit p if there is no n-tuple a1, …, an ∈ |A| which realizes p.

We have already seen how to use the compactness theorem to construct a model, which realizes p. It is somewhat more difficult to construct a model, which omits p. Indeed, such a model may not even exist. The omitting types theorem gives a sufficient condition for the existence of a model of T, which omits p. (If T is complete, this sufficient condition is also necessary.)

Definition 4.17 We say that p is principal over T if it is generated by a single formula, i.e. there exists $\varphi \in \Phi n$ (T) such that $p = \{\psi \in \Phi n\ (T)$: $T \models \forall v1 \cdots \forall vn(\varphi(v1, \ldots, vn) \rightarrow \psi(v1, \ldots, vn))$. Such a φ is called a generator of p.

We say that p is essentially nonprincipal over T if p is not included in any principal n-type over T.

Theorem 4.9 (the omitting types theorem) Let T be a countable theory and let p be an n-type over T. Suppose that p is essentially nonprincipal over T. Then there exists a model of T, which omits p.

Note that the hypothesis that T is countable cannot be omitted from the theorem.

Proof This proof is by the well-known Henkin techniques, which results in a model all of whose elements are the interpretations of constant symbols from some language. The construction for this proof in particular is done so that the theory this model satisfies asserts that each tuple of constants fails to satisfy at least one f from each type pi. As a result, the model will not realize any of the types.

Standard example is considering a language with one binary connective, $\in$.

Let M be the model $\omega \in \omega$, the ordinal standard well ordering. Let T denote the theory of this model. The set of formulas $p(x)$: $\{n \in x |\ n \in {}_{\omega}\omega\}$ is a type. Let $p_0 \in p(x)$. Take the successor of the largest ordinal mentioned in the set of formulas $p_0(x)$. Then this will clearly contain all the ordinals mentioned in $p_0(x)$.Thus we have that $p(x)$ is a type. Now, note that $p(x)$ is not realized in M. Since that would imply there is some $n \in \omega$ that contains every element of w.

4.9 POSITIVE CATEGORIES AND CONSISTENCY MODELS

Define the category $L_{P,\omega}$ to be the category with objects be the positive fragment obtained from Kiesler fragment and arrows the subformula ordering on $L_{P,\omega}$ preorder on formulas.

Define a functor F: $\mathbf{L}^{Op}_{\mathbf{P},\omega} \rightarrow$ **Set** by a list of sets Mn and functions fn. The functor F is a list of

$L_{P,\omega}$ sets Fn, consisting of:

(a) the sets corresponding to an initial structure on L, for example the free syntax tree on $L_{P,\omega}$ structure, where to f(t1, t2,.. tn) in L there corresponds the equality relation f(t1, …, tn)=ft1, …, tn in Set;.

(b) the functions fi:Fi+1 $\rightarrow$ Fi on algebras.

Theorem 4.10 (Infinitary Positive Omitting Fragments)

Let $T<i, i+1>$ be the complete theory in $L<i, i+1>$ = Li intersect Li+1, defined by Th (A(Fi) intersect A(Fi+1)), Let Ti and Ti+1 be arbitrary consistent positive theories for the subfragments Li and Li+1, respectively, satisfying Ti contains $T<i, i+1>$ and Ti+1 contains $T<i, i+1>$. Let Ai be A(Fi) and Bi be A(Fi+1).

(i) There are iterated elementary extensions Bi+1> Bi and an embedding fi: Ai < Bi

(ii) Slalom between the language pairs Li, Li+1 gates to a limit model for L, a model M for $L_{P,\omega}$.

Proof (Similar to 4.7)

Recall that A consistent theory is preserved under homomorphism's iff T has a set of positive axioms.

Positive forcing had defined T* to be T augmented with induction schemas on the Generic diagram functions. That can effectively generates Tarskian models since Tarskian presentations can be created with Skolemization on arbitrary sentences on generic diagrams with the Skolem functions instantiating the generic diagram functions.

Proposition 4.7 Let R and D be models for L. Then R is isomorphically embedded in D iff D can be expanded to a model of the diagram of R.

Proposition 4.8 Let R and D be models for L. Then R is homomorphically embedded in D iff D can be expanded to a model of the positive diagram of R.

Let Σ be a set of formulas in the variables x1, ..., xn. Let R be a model for L. We say that R realizes Σ iff some n-tuple of elements of A satisfies Σ in R. R omits Σ iff R does not realize Σ.

For our purposes we define a new realizability basis.

Consider a complete theory T in L. A formula φ (x1, ..., xn) is said to be complete (in T) iff for every formula ψ(x1, ..., xn), exactly one of T ⊨ φ ψ or T ⊨ φ¬ ψ. A formula θ (x1, ..., xn) is said to be completable (in T) iff there is a complete formula φ (x1, ..., xn) with T models φ θ. If that can't be done θ is said to be incompletable.

Theorem 4.11 Let L1, L2 be two positive languages. Let L = L1 ∩ L2. Suppose T is a complete theory in L and T1 ⊃ T, T2 ⊃ T are consistent in L1, L2, respectively. Suppose there is model M definable from a positive diagram in the language L1 ∪ L2 such that there are models M1 and M2 for T1 and T2 where M can be homomorphically embedded in M1 and M2.

T1 ∪ T2 is consistent.

(ii) There is model N for T1 ∪ T2 definable from a positive diagram that homomorphically extends that of M1 and M2.

Theorem 4.12 (author ASL March 07) Let L1, L2 be two positive languages. Let L = L1 ∩ L2. Suppose T is a complete theory in L and T1 ⊃ T, T2 ⊃ T are consistent in L1, L2, respectively. Then

(i) T1 ∪ T2 has a model M, that is a positive end extension on Models M1 and M2 for T1, and T2, respectively

(ii) M is definable from a positive diagram in the language L1 ∪ L2.

Theorem 4.13 Considering a Tarskian presentation P for a theory T that has a positive local realization, with T* we can assert the following. Every formula on the presentation P is completable in T*.

Proof (Chapter 9).

Similar to the above we can define the category $L_{H,\omega}$ to be the category with objects Horn fragments and arrows the subfoumual preorder on formulas.

Define the category $\mathbf{L}_{\mathbf{H},\,\omega}$ to be the category with objects Horn fragments and arrows the subfoumual preorder on formulas.

Define a functor F: $L_{H,\,\omega}^{Op} \rightarrow$ Set by a list of sets Mn and functions fn. The functor F is a list of sets Fn, consisting of:

(a) the sets corresponding to an initial structure on $L_{P,\,\omega}$, for example the free syntax tree structure, where to f(t1, t2,.. tn) in $L_{P,\,\omega}$ there corresponds the equality relation f(t1, …, tn)=ft1, …, tn in Set;

(b) the functions fi:Fi+1 → Fi. ⊆

Proposition 4.9 (Infinitary Horn Fragment Consistency)

Let T<i, i+1> be the complete theory in L<i, i+1> = Li intersect Li+1, defined by Th (A(Fi) intersect A(Fi+1)), Let Ti and Ti+1 be arbitrary consistent positive Horn theories for the subfragments Li and Li+1, respectively, satisfying Ti contains T<i, i+1> and Ti+1 contains T<i, i+1>. Let Ai be A(Fi) and Bi be A(Fi+1). Starting with a basis model A0, A0|L<0,1> and B0|L<0,1> are models of a complete theory, hence, A0|L<0,1> ≡ B0|l<0,1>. It follows that the elementary diagram of A0|l<0,1> is consistent with the elementary diagram of B0|L<0,1>.

(i) There are iterated elementary extensions Bi+1> Bi and an embedding fi: Ai < Bi

(ii) Slalom between the language pairs Li, Li+1 gates to a limit model for L, a model M for $L_{H,\,\omega}$.

The above proof is further developed based on reduced product Horn filers developed in the following chapters.

4.10 MORE ON FRAGMENT CONSISTENCY

Barr based on (Gray, 1989) explores the connection between categories of models of Horn theories and models of finite limit theories that might prove an interesting application for fragment consistent Horn product models for us hindsight.

By an equational Horn theory is meant an equational theory augmented by a set of conditions of the form.

$$[\varphi 1(x) = \psi 1(x)] \wedge \cdots \wedge [\varphi n(x) = \psi n(x)] \Rightarrow [\varphi(x) = \psi(x)]$$

where φ, ψ, φi and ψi are operations in the theory and x stands for an element of a product of sorts to which they all apply.

A generalized equational Horn theory is an equational theory augmented by a set of conditions of the form

[(φi(x) = ψi(x)] ⇒ [φ(x) = ψ(x)], allowing a possibly infinite conjunction.

The following theorems from (Barr, 2000) might be new application areas for fragment consistency models.

Theorem 4.14 The category of models of a universal Horn theory has a generating family of regular projective.

The Horn theory can be taken to be finitary if and only if this generating set can be taken to be finite. This area might be newer application area for fragment consistency. For example, one might consider applying fragment consistent Horn products to address Barr's conjecture that it is possible to characterize the categories of models of a Horn theory by the existence of a family of regular projectives. Discrete posets are a generating family but not a regular generating family, for the category of posets, but not for categories. Since posets are the models of a Horn theory, it might be possible to characterize the categories of models of a Horn theory by the existence of a family of regular generating projectives.

4.11 EXERCISES

1. A theory T is $\aleph_{0-}$categorical if and only if ∀n(H_n(T) is finite. Here, T is assumed to be countable and complete and to have an infinite model.
2. Prove Theorem 5.12.
3. Define a category D<A, G> on model diagrams on for the infinitary fragment $L_{\omega 1\ K}$.
   What role do the generic diagrams play?
4. Let E be a subset of S(I). By the filter generated by E we mean the intersection D of all filters over I which include E: D = ∩ {F: E < F and F is a filer over I}. E is said to have the finite intersection property iff the intersection of any finite number if elements of E is

nonempty. Prove that the filter D generated by E, any subset E of S(I), is a filter over I.

5. Prove Theorem 4.7: The embedding to form elementary chains on Fi’s can be defined by a back and forth model design from the strings language fragments.
6. Prove Theorem 4.5.
7. Prove theorem 4.6.
8. Prove Proposition 4.7 (positive morphic extensions) The following are equivalent:
   (i) Every positive sentence holding on R also holds on |B.
   (ii) There are elementary extensions R < R,’ |B < |B’ such that |B’ is a homomorphic image of R.
9. The category of models of a universal Horn theory has a generating family of regular projectives.
10. Apply fragment consistent Horn products to address Barr’s conjecture that it is possible to characterize the categories of models of a Horn theory by the existence of a family of regular projectives.

KEYWORDS

- **equational theory**
- **functorial models**
- **Infinitary positive language**
- **Robinson’s Consistency Theorem**

CHAPTER 5

ALGEBRAIC THEORIES, CATEGORIES, AND MODELS

CONTENTS

This chapter starts with products on models towards topos characterizations for algebraic theories and models. Basic categorical and functorial algebraic theory models are presented in light of ultraproducts on models.

5.1 ULTRAPRODUCTS ON ALGEBRAS

This section introduces an exposition from Kiesler that is prominent in the ultraproducts developments. The construction is a uniform technique for building models of first order theories with applications in many areas of mathematics. It is attractive because it is algebraic in nature, but preserves all properties expressible in first order logic. The idea goes back to the construction of nonstandard models of arithmetic by Skolem in 1934. In 1948, Hewitt studied ultraproducts of fields. For first order structures in general, the ultraproduct construction was defined by Lo ́s in 1955. The subject developed rapidly beginning in 1958 with a series of abstracts by Frayne, Morel, Scott, and Tarski, which led to the 1962 paper. Other early papers are by Kochen, and by Kiesler. The groundwork for the application of ultraproducts to mathematics was laid in the late 1950s through the 1960s. The purpose of this article is to give a survey of the classical results on ultraproducts of first order structures in order to provide some background for the papers in this volume. Over the years, many generalizations of the ultraproduct construction, as well as applications of ultraproducts to nonfirst order structures, have appeared in the literature.

The cardinality of a set X is denoted by |X|. The cardinality of N is denoted by ω. By the cardinality of A we mean the cardinality of its universe set A. The notation A $\models$ φ(a1,..., an) means that the formula φ(x1,..., xn) is true in A when each xi is interpreted by the corresponding ai. The notation h: A→B means that h is a homomorphism of A into B, that is, h maps A into B and each atomic formula which is true for a tuple in A is true for the h-image of the tuple in B. The notation h: A ⊆ B means that h is an (isomorphic) embedding of A into B, that is, h maps A into B and each quantifier-free formula of L which true for a tuple in A is true for the h image of the tuple in B. h: A~=B means that h is an isomorphism from A onto B, and A ~= B means that A and B are isomorphic

The set of all sentences true in A is called the complete theory of A. A and B are called elementarily equivalent, in symbols $A \equiv B$, if they have the same complete theory. The notation h: $A \prec B$ means that h is an elementary embedding from A into B, that is, h maps A into B and each formula of L which is true for a tuple in A is true for the h-image of the tuple in B. Clearly, h: $A \prec B$ implies that $A \equiv B$. We say that B is an elementary extension of A and write $A \prec B$ if $A \subseteq B$ and the identity map is an elementary embedding of A into B. It is easy to see that if h: $A \prec B$, then B is isomorphic to some elementary extension of A.

A fundamental result in model theory is the compactness theorem:

Theorem 5.1 If every finite subset of a set T of sentences has a model, then T has a model.

Compactness has many applications, with an important one: to construct rich models called saturated models. For a language L, an L-structure A is said to be κ-saturated if every set of first order formulas with fewer than κ parameters from A which is finitely satisfied in A is satisfied in A. A is saturated if it is |A|-saturated. Morley and Vaught [39] proved that any two elementarily equivalent saturated structures of the same cardinality are isomorphic, that each infinite structure A has a saturated elementary extension in each inaccessible cardinal $\kappa \geq |A| + |L|$, and has a κ+-saturated elementary extension of cardinality 2κ whenever $2\kappa \geq |A|$ and $\kappa \geq |L|$.

Given two vocabularies $L1 \subseteq L2$, the reduct of an L2-structure A2 to L1 is the L1-structure A1 obtained by forgetting the interpretation of each symbol of L2 \L1. An expansion of an L1-structure A1 to L2 is an L2-structure formed by adding interpretations of the symbols of L2 \ L1, that is, an L2-structure whose reduct to L1 is A1.

Let S is a set of formulas and/or terms, the signature of S is the set of all nonlogical symbols occurring in it. This is sometimes called in the literature the similarity type of S. Note that = never belongs to the signature since it is a logical symbol. A sentence is a formula with no free variables.

Definition 5.1 A theory T is complete if for all sentences σ, either $\sigma \in T$ or $\neg\sigma \in T$, provided σ has signature ⊆ T's signature.

Examples: The theory of groups is not complete. The theory of fields of characteristic 0 is not complete (e.g., $\exists x(x \cdot x = 1 + 1)$ is true in R, false in Q). The theory of algebraically closed fields of characteristic 0 is complete.

Definition 5.2 Two structures A and B are elementarily equivalent (written A ≡ B) if they have the same signature and satisfy the same sentences. In other words, Th(A) = Th(B).

Proposition 5.1 For a theory T the following are equivalent.

1. T is complete;
2. all models of T are elementarily equivalent;

We use the following notational convention: φ(x1,..., xn) denotes a formula φ whose free variables are among x1,..., xn.

A theory T is said to admit elimination of quantifiers if for all formulas φ(x1, …, xn), n ≥ 1, there exists a quantifier free formula φ∗(x1, …, xn) such that:

T ⊨ ∀x1 …∀xn(φ(x1, …, xn) ↔ φ∗(x1, …, xn)).

Vaught's test is another method for establishing completeness of theories. It is less versatile than quantifier elimination, but much easier to use. Two structures A and B are said to be isomorphic (written A ≡ B) if signature (A) = signature (B) and there exists an isomorphic map of A onto B, that is, h: |A| → |B| such that

1. h is one-one and onto;
2. R_A(a1, …, an) if and only if R_B(a1, …, an);
3. h(s_A(a1, …, an))= s_B (i(a1), …, i(an));
4. h(c_A) = c_B.

Note that isomorphic structures are "essentially identical." In particular they satisfy the same sentences, that is, A ; B implies A ≡ B.

The cardinality or power of a structure A is the cardinality of its universe. For instance, we say that A is countable if and only if |A| is countable, etc. The power of A is denoted ||A|| or card(A).

A theory T is said to be κ-categorical if (i) T has at least one model of power κ; (ii) any two such models are isomorphic. Further specifics might be explored on ultraproducts at Ekloff (1977).

5.2 ULTRAPRODUCTS AND ULTRAFILTERS

Following Kiesler's, model theory exposition, the cardinality of a set X is denoted by |X|. The cardinality of N is denoted by ω. The set of all subsets of a set I is denoted by P(I), and the set of finite subsets of I by Pω(I). Given mappings f : X → Y and g: Y → Z, the composition g ∘ f: X → Z is the mapping $x \rightarrow$ g(f(x)). By the cardinality of A we mean the cardinality of its universe set A. The notation A ⊨ φ(a1,..., an) means that the formula φ(x1,..., xn) is true in A when each xi is is assigned the corresponding ai. The notation h: A→B means that h is a homomorphism of A into B, that is, h maps A into B and each atomic formula which is true for a tuple in A is true for the h-image of the tuple in B. The notation h: A ⊆ B means that h is an (isomorphic) embedding of A into B, that is, h maps A into B and each quantifier-free formula of L which is true for a tuple in A is true for the h-image of the tuple in B.

h : A ; B means that h is an isomorphism from A onto B, and A ; B means that A and B are isomorphic. The set of all sentences true in A is called the complete theory of A. A and B are called elementarily equivalent, in symbols A ≡ B, if they have the same complete theory. The notation h : A ≺ B means that h is an elementary embedding from A into B, that is, h maps A into B and each formula of L which is true for a tuple in A is true for the h-image of the tuple in B. Clearly, h : A ≺ B implies that A ≡ B. We say that B is an elementary extension of A and write A ≺ B if A ⊆ B and the identity map is an elementary embedding of A into B. It is easy to see that if h : A ≺ B then B is isomorphic to some elementary extension of A.

A fundamental result is the compactness theorem, which says that if every finite subset of a set T of sentences has a model, and then T has a model. One application of compactness is the construction of rich models called saturated models. Saturated models prove essential in this author's forthcoming chapters treatments on models on topoi. An L-structure A is said to be κ-saturated, if every set of first order formulas with fewer than κ parameters that is finitely satisfied in A is satisfied in A. A is saturated if it is |A|-saturated. Morley and Vaught proved that any two elementarily equivalent saturated structures of the same cardinality are isomorphic, that each infinite structure A has a saturated elementary extension in each inaccessible cardinal $\kappa \geq |A| + |L|$, and has a κ+-saturated elementary extension of cardinality 2κ whenever $2\kappa \geq |A|$ and $\kappa \geq |L|$.

Given two vocabularies $L1 \subseteq L2$, the reduct of an L2-structure A2 to L1 is the L1-structure A1 obtained by forgetting the interpretation of each symbol of $L2 \setminus L1$. An expansion of an L1-structure A1 to L2 is an L2-structure formed by adding interpretations of the symbols of $L2 \setminus L1$, that is, an L2-structure whose reduct to L1 is A1.

Definition 5.3 Let I be a non-empty set. A proper filter U over I is a set of subsets of I such that:

(i) U is closed under supersets; if $X \in U$ and $X \subseteq Y \subseteq I$ then $Y \in U$.
(ii) U is closed under finite intersections; if $X \in U$ and $Y \in U$ then $X \cap Y \in U$.
(iii) $I \in U$ but $\emptyset \in/U$.

An ultrafilter over I is a proper filter U over I such that: (iv) For each $X \subseteq I$, exactly one of the sets $X, I \setminus X$ belongs to U.

Theorem 5.2 (Tarski) Every proper filter over a set I can be extended to an ultrafilter over I.

The only ultrafilters over I which are not free are the principal ultrafilters. The principals are of the form $U = \{X \subseteq I : i0 \in X\}$ for some $i0 \in I$. For a set I of finite cardinality n, every ultrafilter over I is principal, and thus there are only n ultrafilters over I. Let us now define the **ultraproduct** operation on **sets**. Let U be an ultrafilter over I, and for each $i \in I$ let Ai be a nonempty set. The ultraproduct Π_U Ai is obtained by first taking the cartesian product $\Pi_{i \in I}$ Ai and then identifying two elements which are equal for U-almost all $i \in I$.

Definition 5.4 Let U be an ultrafilter over I. Two elements f,g of the Cartesian product $i \in I$ Ai are said to be U-equivalent, in symbols $f =_U g$, if the set $\{i : f(i) = g(i)\}$ belongs to U. The U-equivalence class of f is the set $f_U = \{g : f =_U g\}$.

The ultraproduct Π_U Ai is defined as the set of U-equivalence classes $\Pi_{i \in I} Ai = \{f_U : f \in \Pi Ai\}$.

U $i \in I$ embedding is the mapping $d : A \rightarrow \Pi_U A$ such that d(a) is the U-equivalence class of the constant function with value a. It is easily seen that d is injective.

In the above definition, it is easily checked that $=_U$ is an equivalence relation on $\Pi_{i\in I}$Ai. Given a nonempty set A, the **ultrapower** of A modulo U is the defined i∈I as $\Pi_U A = \Pi_U Ai$ where Ai =A for each I ∈ I.

The **ultraproduct** operation on **first order structures** is defined as follows. For each i ∈ I, let Ai be an L-structure with universe set Ai. Briefly, the ultraproduct Π_U Ai is the unique L-structure with universe Π_U Ai such that each basic formula holds in the ultraproduct if and only if it holds in Ai for U-almost all i. Here is the formal definition.

Definition 5.5 Given an ultrafilter U over I and L-structures Ai,i ∈ I, the ultraproduct Π_U Ai is the unique L-structure B such that:

The universe of Bi s the set B= Π_UAi For each atomic formula φ(x1, ..., xk) which has at most one symbol from

the vocabulary L, and each f1,...,fk ∈ $\Pi_{i\in I}$ Ai, B ⊧ φ($f1_U$,...,fk_U) iff {i : A |= φ(f1(i),...,fk(i))} ∈ U.

Using the properties of ultrafilters, one can check that there is a unique L- structure B with the above properties, so the ultraproduct is well-defined. As with sets, the ultrapower of an L-structure A modulo U is defined as the ultraproduct Π_U A = Π_U Ai where Ai Í A for each i ∈ I.

A fundamental theorem of Lo ́s, that makes ultraproducts useful in model theory. It shows that a formula holds in an ultraproduct U Ai if and only if it holds in Ai for U-almost all i.

Theorem 5.3 (Lós) Let U be an ultralfilter over I, and let Ai be an L-structure for each i ∈ I. Then for each formula φ(x1,...,xn) of L and each f1,...,fn ∈ $\Pi_{i\in I}$Ai , we have i∈I, Π_U Ai ⊧ φ($f1_U$,...,fn_U) iff {i : Ai ⊧ φ(f1(i),...,fn(i))} ∈ U.

Corollary 5.1 For each set of sentences T in L, every ultraproduct of models of T is a model of T.

Corollary 5.2 For each L-structure A and ultrafilter U over I, d : A ≺ $\Pi_U A_i$

If A is finite, d:A ;P $_U$ A.

In applications, it is often convenient to rename the elements of an ultrapower. We say that an isomorphic embedding h : A → B is an ultrapower embedding if h=j∘d for some isomorphism j: Π_U A $\simeq$ B. The natural embedding d:A→ Π_U A is an ultrapower embedding. We say that B is an ultrapower extension of A if A $\subseteq$ B and the identity map ι : A → B is an ultrapower embedding.

An important property of the ultraproduct construction is that it behaves well when new symbols are added to the vocabulary.

Proposition 5.2 (Expansion Property) Suppose L1 $\subseteq$ L2, and for each i ∈ I, Ai is an L1-structure and Bi is an expansion of Ai to L2. Then for every ultrafilter U over I, Π_U Bi is an expansion of Π_U Ai.

Given an ultrafilter U over as et I and a mapping h:A→B, define f_U to be the mapping f_U Π_U → $(h\circ f)_U$ from U_A into U_B.

Proposition 5.3 Let U be an ultrafilter over I. The mapping h → Π_U h is a functor on the category of all homomorphisms h : A → B between L-structures. If h:A →B then h: Π_UA → Π_U B. If h is surjective, then so isΠ_U h. If h is an isomorphic embedding, then so is Π_U h. If h is an elementary embedding, then so is Π_U h.

The initial interest in ultraproducts in the late 1950's was sparked by the discovery of a proof of the Compactness Theorem for first order logic via ultraproducts. This proof was attractive because it gave a direct algebraic construction of the required model.

Theorem 5.4 (Ultraproduct Compactness) Let S be an infinite set of sentences of L and let I be the set of all finite subsets of S. For each i∈I let Ai be a model of i. Then there is an ultrafilter U over I such that the ultraproduct U Ai is a model of S.

Proof (Kiesler) For each i ∈ I, let Xi be the set of all j ∈ I such that i $\subseteq$ j. Let F be the set of all X$\subseteq$I such that X$\supseteq X_i$ for some i∈I. Notet hat i∈X_i,and $X_i \cup_j = X_i \cap X_j$. It follows that F is a proper filter over I. By Theorem above, F can be extended to a nultralfilter U over I. For each φ∈S and j ∈ X{φ}, A_j is a model of φ. Moreover, X{φ}∈U. Therefore by Lo ́s' Theorem, UA_i is a model ofφ.Hence UA_i is a model of S.

The compactness theorem is an easy corollary of this result. For this reason, the ultraproduct construction can be used as a substitute for the compactness theorem. Another important property of ultraproducts is that an ultraproduct of ultraproducts is isomorphic to a single ultraproduct.

5.3 ULTRAPRODUCT APPLICATIONS TO HORN CATEGORIES

Algebraic operations on the Cartesian product $\Pi_i M_i$ are defined in the usual way (for example, for a binary function +, $(a + b)\, i = ai + bi$), and an equivalence relation is defined by $a \sim b$ if and only if {i ∈ I: ai = bi} ∈ U.

Let I be a nonempty set. Let S(I) be the set of all subsets of I. A filter D over I is defined to be a set D < S(I) such that I ∈ D; if X, Y ∈D, then X ∩ Y ∈D; if X ∈D and X<Z<I, then Z ∈D. Note that every filter D is a nonempty set since I ∈D. example filters are the trivial filter D = {I}. The improper filter D = S(I).

For each Y < I, the filter D = {X <I; Y < X}; this filter is called the principal filter generated by Y. D is said to be a proper filter iff it is not the improper filter S(I).

Theorem 5.6 (Horn Consistency) Let L1, L2 be two Horn languages. Let L = L1 ∩ L2. Suppose T is a complete theory in L and T1 ⊃ T, T2 ⊃ T are consistent in L1, L2, respectively. Then T1 ∪ T2 has a model M that is definable from a generic (finite) direct product diagram on the language L1 ∪ L2.

Amongst the few techniques to design models is the ultraproduct technique that is further developed above and in part applied here.

Let I be a nonempty set. Let S(I) be the set of all subsets of I. A filter D over I is defined to be a set D < S(I) such that I ∈ D; if X,Y ∈ D, then X∩Y ∈ D; If X ∈D and X<Z<I, then Z εD. Note that every filter D is a nonempty set since I ε D. example filters are the trivial filter D = {I}. The improper filter D = S(I). For each Y < I, there is a filter D = {X <I; Y < X}; this filter is called the principal filter generated by Y. D is said to be a proper filter iff it is not the improper filter S(I).

Let E be a subset of S(I). By the filter generated by E we mean the intersection D of all filters over I which include E: D = ∩ {F: E < F and F is a filer over I}. E is said to have the finite intersection property iff the intersection of any finite number if elements of E is nonempty.

Can prove that the filter D generated by E, any subset E of S(I), is a filter over I. Now let us consider models with reduced products. Let I be a nonempty set, let D be a proper filter over I, for each I in I let Ri be a model for L. applying the convention that in Ai the relation symbols P are interpreted by Ri, functions F by Gi and constants c by ai.

The **reduced product** Π_DRi is the model for L defined as follows:

(i) the universe set of Ri is Π_D Ai.

(ii) Let p be an n-placed relation symbol of L. The interpretation of p in Π_D Ai is the relation

S such that $S(f1_D...fn_D)$ iff $\{i\in I: Ri(f(i)...fn(i))\} \in D$. (iii) Let F be an n-placed function symbol of L. Then F is interpreted in ΠD Ai by the function H given by $H(f1_D...fn_D) = <Gi(f1(i)....fn(i)): i \in I >D$. (iv) Let c be a constant of L. Then c is interpreted by the element $b \in \Pi_D$ Ai, where $b <ai : i \in I >D$

The definitions can be proved consistent. From here there are important theorems called expansion theorem and ultraproduct theorem obtained. We only apply the expansion theorem paraphrased: that we can expand a language and take reduced products such the new reduced product is an expansion of the reduced product for the starting language.

The definitions can be proved consistent. From here there are important theorems called expansion theorem and ultraproduct theorem obtained. We only apply the expansion theorem paraphrased: that we can expand a language and take reduced products such the new reduced product is an expansion of the reduced product for the starting language. The filtering techniques the author has developed since positive forcing, here, and specifics (Nourani, 2005b, 2006) are applied to create models for the arbitrary fragments, for example Horn fragments to obtain the ultraproducts.

Proposition 5.5 (i) Let φ (x1…xn) be a Horn formula and let Ri, $i \in I$ be models for language L. Let D be a proper filter over I, and let a1….an ∈ ii∈ i.

If $\{i \in I: Ri \models \varphi[(a1(i) \ldots an(i)]\} \in D$, then

(a) Π_D Ri $\models \varphi[a1_D \ldots an_D]$

(b) Every Horn sentence is preserved under reduced products.

(c) Every Horn sentence is preserved under direct (finite or arbitrary) products.

The following propositions are from basic model theory, for example, Chang-Kiesler.

Proposition 5.6 Let φ be a universal sentence. Then φ is a (finite) direct product sentence iff φ is equivalent to a universal Horn sentence.

Proposition 5.7 Let φ be a universal sentence. Then φ is a (finite) direct product sentence if and only if φ is equivalent to a universal Horn sentence.

Proposition 5.8 Let φ be an existential sentence. Then φ is a (finite) direct product sentence if φ is equivalent to an existential Horn sentence.

Define the category $L_{H,\omega}$ to be the category with objects Horn fragments and arrows the subfoumual preorder on formulas.

Define a functor F: $L_{H,\omega}^{Op}$ → Set by a list of sets Mn and functions fn. The functor F is a list of sets Fn, consisting of:

(a) the sets corresponding to an initial structure on $L_{P,\omega}$, for example the free syntax tree structure, where to f(t1, t2,...tn) in $L_{P,\omega}$ there corresponds the equality relation f(t1, …, tn)=ft1…tn in Set;
(b) the functions fi : Fi+1 → Fi.

Theorem 5.5 (Infinitary Horn Fragment Consistency) Define the category $L_{H,\omega}$ to be the category with objects Horn fragments and arrows the subformula preorder on formulas.

Let T<i, i+1> be the complete theory in L<i, i+1> = Li intersect Li+1, defined by Th (A(Fi) intersect A(Fi+1)), Let Ti and Ti+1 be arbitrary consistent positive theories for the subfragments Li and Li+1, respectively, satisfying Ti contains T<i, i+1> and Ti+1 contains T<i, i+1>. Let Ai be A(Fi) and Bi be A(Fi+1). Starting with a basis model A_0, A_0|L<0,1> and B_0|L<0,1> are models of a complete theory, hence, A_0|L<0,1> ≡ B_0|l<0,1>. It follows that the elementary diagram of A_0|l<0,1> is consistent with the elementary diagram of B_0|L<0,1>.

(i) There are iterated elementary extensions Bi+1> Bi and an embedding fi: Ai < Bi
(ii) Slalom between the language pairs Li, Li+1 gates to a limit model for L, a model M for $L_{P,\omega}$.

Proof Outline

Considering Chapter 4 we carry on with Horn direct product models on new applications from the above Propositions 5.8. The significance of filters and ultraproducts are on a glimpse on an infinite consistency area on Horn models, where the standard "back and forth" techniques on infinitary logic were lifted to infinite paring on language fragments in general since the authors 1997. Horn instance is here on the proof outline where the limits rely on the product filters. Further specifics on an alternate chapter on models.

5.4 ALGEBRAIC THEORIES AND TOPOS MODELS

Informally in mathematical logic, an *algebraic theory* is one that uses axioms stated entirely in terms of equations between terms with free variables. Inequalities and quantifiers are specifically disallowed. Sentential logic is the subset of first-order logic involving only algebraic sentences.

An algebraic theory consists of a collection of n-ary functional terms with additional rules (axioms).

E.g., a group theory is an algebraic theory because it has two functional terms, a binary operation $a * b$ a nullary operation 1 (neutral element), and a unary operation $x \rightarrow x^{-1}$ with the rules of associativity, neutrality and inversion.

This is opposed to geometric theory, which involves partial functions (or binary relationships) or existential quantifiers, see for example, Euclidean geometry where the existence of points or lines is postulated.

An Algebraic Theory **T** is a category whose objects are natural numbers 0, 1, 2, …, and which, for each n, has an n-tuple of morphisms:

$proj_i: n \rightarrow 1, i = 1, \ldots, n$

This allows interpreting n as a Cartesian product of n copies of 1.

Example. Let's define an algebraic theory **T** taking hom(n, m) = m-tuples of polynomials of n free variables $X_1, \ldots, X_n$ with integer quotients and with substitution as composition. In this case $proj_i$ is the same as X_i. This theory T is called the theory of commutative rings.

In an algebraic theory, any morphism $n \rightarrow m$ can be described as m morphisms of signature $n \rightarrow 1$. These latter morphisms are called n-ary *operations* of the theory.

If E is a category with finite Cartesian products, the full subcategory Alg(**T**, E) of the category of functors [**T**, E] consisting of those functors that preserve finite products is called *the category of* **T**-*models* or **T**-*algebras*.

Note that for the case of operation $2 \rightarrow 1$, the appropriate algebra A will define a morphism $A(2) \approx A(1) \times A(1) \rightarrow A(1)$

5.5 FREE THEORIES AND FACTOR THEORIES

Algebraic theories were developed by Lawvere (1963) to categorize universal algebra. There are two views to algebraic theories: algebraic theories and categories and algebraic theories as functors. The specific presentation for algebraic theories as categories is dual to Lawever's original, following a concrete treatise by ADJ (1975) since the authors carried on his doctoral dissertation with the ADJ group: University of California and IBM Mathematical Sciences research a decade from that. An axiomatic route is Elgot (1974). The techniques is to define a free category then taking a factor, similar to how a group is defined as a factor group from a free group. Like a free group generated by Ω, a set, is a concrete free theory with base Ω— operator domain. A sequence of sets $\Omega = \Omega_0, \Omega_1, \ldots, \Omega_n, \ldots$ is called an operator domain: Ωn is considered a function symbol with n arguments. The concrete free theory with base Ω, is a category with morphism tuples of terms on function symbols from Ω.

Definition 5.6

(1) Let $\Omega = < \Omega_0, \Omega_1, \ldots, \Omega_n .. > = < \Omega n \mid n\ w>$, an w-indexed family of sets, e.g., function symbols. Furthermore. Let $X = <x_1, \ldots, x_n, \ldots>$ $<x_n : n \in w>$ be a w-indexed family of variables. With X and Ω disjoint sets for all n in w. Let Ω denote the set of all function symbols ($\Omega = U\ \Omega n$).

For each n in w, let Tn, the set of n-ary Ω-terms, be the smallest subset of ($\Omega\ U$ Xn U {(,), })* satisfying: Xn subset Tn.

(2) If a is in Ωp and $<t_0, \ldots, t_{p-1}>$ in T_n.
Note that $T^\circ{}_n = \{\ \}$, the empty string, so that a() is in Tn for every a in Ω_0.

A concrete free theory with base Ω is the category *Ñ[Ω]* with objects | Ñ[Ω |the ordinals {0,1, ….} = w and morphisms for all n, p in w Ñ[Ω] (n.p) is the set of all triples <n, t, p> such that t = (t0, t1, …, tn–1) ti in Tp for each i<n, that is, t is in $T^n{}_p$.

t is an n-tuple of p-ary Ω-terms. Write t: n → p for <n, t, p>. Composition is substitution, that is given a = (t_0, …., t_{n-1}) and b = (t_0,' …,t'_{m-1}) such that

a : n →:m and b = n → p, define a b = (t''_0, …, t''_{n-1}): n →p. For each n 1n (x_0, …, x_{n-1}) is the identity morphism for n.ti" is obtained form simulates variable replacement t'j for tj.

5.6 T-ALGEBRAS AND ADJUNCTIONS

5.6.1 LIMITS, COLIMITS, AND CONES

Definition 5.7 Let *F*: *J* → *C* be a diagram in *C*. Formally, a diagram is nothing more than a functor from *J* to *C*. The change in terminology reflects the fact that we think of *F* as indexing a family of objects and morphisms in *C*. The category *J* is thought of as an "index category". One should consider this in analogy with the concept of an indexed family of objects in set theory. The primary difference is that here we have morphisms as well.

Let *N* be an object of *C*. A **cone** from *N* to *F* is a family of morphisms

Y (X): N → F(X),

for each object *X* of *J* such that for every morphism *f*: *X* → *Y* in *J* the following diagram commutes:

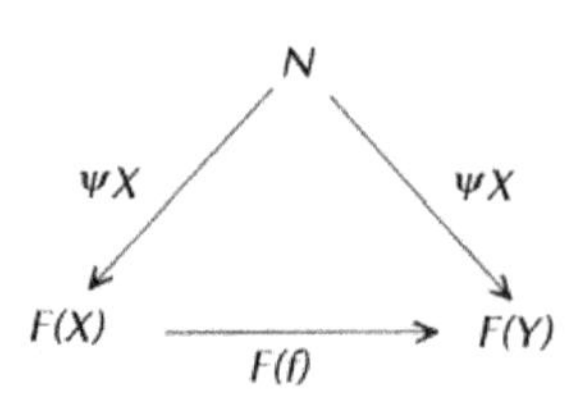

The (usually infinite) collection of all these triangles can be (partially) depicted in the shape of a cone with the apex *N*. The cone ψ is sometimes said to have **vertex** *N* and **base** *F*.

One can also define the dual notion of a **cone** from *F* to *N* (also called a **cocone**) by reversing all the arrows above. Explicitly, a cone from *F* to *N* is a family of morphisms.

ψ_X: F(X) → N

for each object *X* of *J* such that for every morphism *f*: *X* → *Y* in *J* the following diagram commutes:

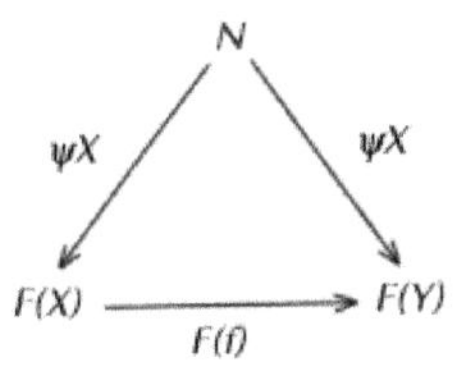

Alternative statements are as follows.

Cones are maps from an *object* to a *functor* (or vice-versa). We can define them as morphisms or objects in some suitable category.

Let *J* be a small category and let *CJ* be the category of diagrams of type *J* in *C* (this is nothing more than a functor category). Define the diagonal functor Δ: *C* → *CJ* as follows: Δ(*N*): *J* → *C* is the constant functor to *N* for all *N* in *C*.

If *F* is a diagram of type *J* in *C*, the following statements are equivalent:

- ψ is a cone from *N* to *F*
- ψ is a natural transformation from Δ(*N*) to *F*
- (*N*, ψ) is an object in the comma category (Δ ↓ *F*)

The dual statements are also equivalent:

- ψ is a cocone from *F* to *N*
- ψ is a natural transformation from *F* to Δ(*N*)
- (*N*, ψ) is an object in the comma category (*F* ↓ Δ)

These statements can all be verified by a straightforward application of the definitions. Thinking of cones as natural transformations we see that they are just morphisms in CJ with source (or target) a constant functor. Now we can define the *category of cones to F* as the comma category ($\Delta \downarrow F$). Morphisms of cones are then just morphisms in this category. As one might expect a morphism from a cone (N, ψ) to a cone (L, φ) is just a morphism $N \rightarrow L$ such that all the "obvious" diagrams commute (see the first diagram in the next section).

Likewise, the **category of cocones from *F*** is the comma category ($F \downarrow \Delta$).

Limits and colimits are defined as **universal cones**. That is, cones through which all other cones factor. A cone φ from L to F is a universal cone if for any other cone ψ from N to F there is a unique morphism from ψ to φ.

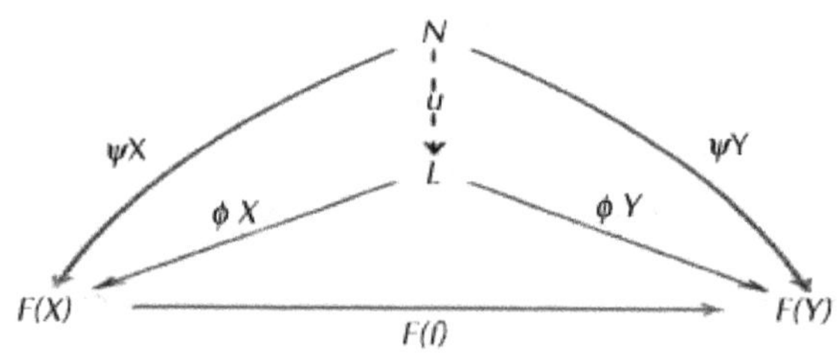

Equivalently, a universal cone to F is a universal morphism from Δ to F (thought of as an object in CJ), or a terminal object in ($\Delta \downarrow F$).

Dually, a cone φ from F to L is a universal cone if for any other cone ψ from F to N there is a unique morphism from φ to ψ.

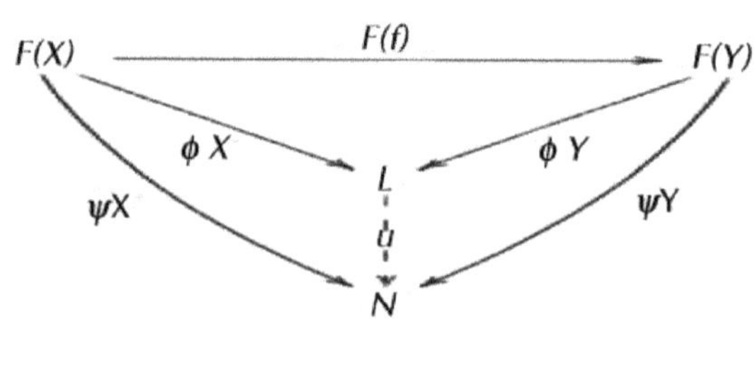

Equivalently, a universal cone from F is a universal morphism from F to Δ, or an initial object in $(F \downarrow \Delta)$.

The limit of F is a universal cone to F, and the colimit is a universal cone from F. As with all universal constructions, universal cones are not guaranteed to exist for all diagrams F, but if they do exist they are unique up to a unique isomorphism.

5.7 THEORY MORPHISMS, PRODUCTS AND CO-PRODUCTS

On this section we further explore a category-based model-theoretic characterization based on algebraic theories.

A theory congruence on **T** is a family of congruence relations on the family of sets $<\mathbf{T}(n, m \mid n, m \in \mathbf{w}>$ which is compatible with the structure of **T.**

Definition 5.8 Given a concrete free theory **T** and a theory congruence **K** on **T, the** factor theory of **T** over **K**, denoted by **T/K**, to be the category with objects $\mid T/K| = |T| = \mathbf{w}$:

Morphisms: **(T/K)** (n, m) **is the set of** $\mathbf{K}_{n,\,m}$ -equivalence classes of **T**(n.m).

Composition is by representatives: for each a $\in$ T(n, m), write $\alpha\sim$ to denote its K n, m equivalence class. Then foe all $\alpha i\sim\ \in T(ni, n\, i+1)$, i=0,1, … $\alpha 0\sim\alpha 1\sim = \alpha 0 \alpha i\sim$.

Definition 5.9 A theory congruence K on a concrete free theory **T** is a bi-indexed family of sets $k = <Kn, m \mid n, m \in w>$ such that for all n, m and p:

(1) Kn, m $\subseteq$ **T** (n, m);
(2) Kn, m is an equivalence relation on **T**(n, m).
(3) If $<\alpha 0, \alpha 1> \in$ Kn, m and $<\beta 0, \beta 1> \in$ Km, p then $<\alpha 0\beta 0, \alpha 1\beta 1> \in$ Kn, p;
(4) If $<\alpha i, \beta i> \in$ K1, p for all i $\in$ n then $<(\alpha 0, \ldots, \alpha n\text{-}1), (\beta 0, \ldots, \beta n\text{-}1)> \in$ Kn, p.

Definition 5.10 Given a concrete free theory **T** and the theory congruence on **T**, define the factor theory of **T** over K, denoted **T**/K to be the category with objects $|\mathbf{T}/K| = |\mathbf{T}| = \omega$

Morphisms: (**T**/K) (n.m) is the set of Kn, m-equivalence cases of **T**(n, m);

Composition is by representatives for each a ∈ **T**(n, m) write <a> to denote its Kn, m-equivalence class. Then for all <αi> ∈ **T**(ni, ni+1), i=0,1, we have <α0><α1> =<α0α1>.

Definition 5.11 By an algebraic theory T we mean a category **T**, which is isomorphic to a factor theory of a concrete free theory.

Nonnegative integers are usually used to represent objects of an arbitrary algebraic theory (see Exercise 3).

Replacing concrete free theory by "algebraic theory" we can extend the concepts of theory congruence and a factor theory to arbitrary algebraic theories.

For example, if **T** is an algebraic theory and K is a theory congruence on **T**, then **T/K** is an algebraic theory.

Theorem 5.6 Let T be an algebraic theory and let a, b ∈ **T** (n, m), then a= b iff for all i<n, xi a = xi b (xi = (xi) 1→n.)

Proof Exercise 4.

Example theory of groups: Let W = {e}, {–1}, {.}, Æ, …., Æ, … and let **N** [Ω] be the free theory with base W. Write *x*–1 for −1(*x*) and *x*.*y* for. (*x*, *y*) then W-terms are exactly the expressions that are used to describe groups. Now let K be the smallest theory congruence on **N** [Ω] such that <e.x0, x0>, <*x*-1.x0, e>, <x0.e, x0>, <x0.x0–1, e> ∈ K1,1 and n <x0.(x1, x2), (x0.x1).x2> ∈ K1,3. Then we call **N** [Ω] /K the *theory of groups.*

5.8 ALGEBRAS AND THE CATEGORY OF ALGEBRAIC THEORIES

Example categories are category of theories.

Objects are theories **T**

Morphisms **f: T → T'** are functors F such that

1. F(xi) = xi
2. F((f0, …., fn+1)) = [F(f0, …, F(fn-10))]
3. |F| (n) = n, n ∈ w.

Now define a category on signature theories **Sig**

Objects are sequences Σ0, Σ1, Σ2, ….
Morphisms S → S‘ are sequences fi: S → S,’ i ∈ w.
The functor **Th** ⇒ **Sig** can be defined as follows:

Definition 5.12 Category of algebraic theories, denoted **Th,** is defined as follows:

Objects: |Th| = the calss of all algebraic theories;
Morphisms: Given theories T1 and T2 define **Th(T1, T2)** be the set of all functors F: T1 → T2 that are identity on objects, |F| = 1w;
(i)xiF = xi for all xi ∈ *x*
(ii)for all αi ∈ T1 (1, p), I <n, $(\alpha_0, \ldots, \alpha_{n-1})F = (\alpha_0 F, \ldots, \alpha_{n-1}F)$.
The morphisms are called theory morphisms. There are basics to note. Let R = <Rn, m | n, m ∈ w >, where Rn, m ⊆ **T**n, m *x* **T** n, m for all n, m ∈ w. Then there exists a smallest congruence K = <Kn, m | n, m ∈ w> on Rn, m ∈ Kn, m, for all n, m ∈ w.

5.8.1 T-ALGEBRAS AND PROJECTIONS

Consider the category of categories Cat, where functors server as the morphisms. Functors, however, can serve as the objects of a category. That is presented in this section in brief following ADJ 1976. Let us define T-Algebras based on projections. Let A be a set: for each n, and i <n, let $p_I{}^n$: An → A denote the ith projection from $A^n \rightarrow A$, that is each <a10, …, an-1> A^n, $(a_0, \ldots, a_{n-1})\, p_I{}^n = ai$.

Definition 5.13 Let ***T*** be an algebraic theory. A T-Algebra is a functor A: $\boldsymbol{T}^{op} \rightarrow$ **Set** such that:

(i) For each n >=0, n |A| = $(1\,|A|)^n$, the n-fold Cartesian power of 1 |A|
If x1: 1 → n, in ***T*** then $x_I{}^{op} = p_i{}^{n:}$ (1|A|) n → 1 |A|.
Set 1 |A| is the carrier of the algebra A. In general, we write A to denote the carrier as well as the algebra.

Now we carry on to examples of species and then to instantiate natural transformations on T. Description of species of algebras corresponds to an algebraic theory, e.g., the theory ***N***[Ω] /K where K is the smallest theory congruence on ***N***[Ω], containing E, a set of educational axioms. Conversely, given a theory T, each presentation of it as a factor theory $N[\Omega]$ /K yields a corresponding description of a species with ranked alphabet W and axioms K-preceding sections. Now let us define the category of ***T***-Algebras. First define the morphisms as follows.

Definition 5.14 Given T-algebras A, B: $\mathbf{T}^{op} \to$ **Set,** define **a T**-homomorphism from A to B with the mapping h: $A \to B$ from the carrier of A to the carrier of B such that for each $a \in \boldsymbol{T}(1, n)$ and all $<a_0, \ldots, a_{n-1}> \in A^{n,}$ $(a_0, \ldots, a_{n-1})(\alpha^{op} A)$

$h = (a_0, \ldots, a_{n-1}h)((\alpha^{op} B)$.

Definition 5.15 Starting with an algebraic theory **T,** define the category of T-algebras, denoted by **T^** as the category with | **T^** | the class of all T-algebras, where given T-algebras A and B, **T^** (A, B) is the set of all T-homomorphism h from A to B. Given h ∈ **T^** (A, B) and h' ∈ **T^** (A, B), the composite is hh': $A \to C$.

From the above, for example, we can show that (Exercise 9) if we let T be the theory of groups, then **T^** is isomorphic in (**Cat**) to the category of groups.

5.8.2 NATURAL TRANSFORMATIONS ON T

Considering where will applying natural transformations on sheaves for example, on hindsight, let us have a new glimpse to natural transformation since chapter II. Since ***T***-algebras were defined as specific functors, it follows that ***T***-homomorphisms are functor morphisms. We now apply this to generalize to morphisms between arbitrary functors. With common source and target, towards constructing categories with functor objects. Let us consider the notion of a ***T***-homomorphism.

Let A, B: $\boldsymbol{T}^{op} \to$ **Set** be *T*-algebras and let h: $A \to B$ be a ***T***-hommorphism form A to B. For each n >=0 let h_nn: $A_n \to B_n$ be such that for all <a0, …, an-1> h_n = <a0h, a1h, ….an-1h>, in particular we have h1=h. Regarding h as a family of functions h: <$h_{n:}$ An → Bn> | n ∈ w>. That h is a T-homomorphism is conditioned as: For each $\alpha \in \boldsymbol{T}(1, n)$ and all <a0,

…, an–1> ∈ A^n, (a0, …, an-1) (α^{op} A) h_1 = (a0h, a1h, …, an–1) h_n (α^{op} B). More briefly as (α^{op} A) h_1 = h_n (α^{op} B).

Definition 5.16 Let F and G be functors, F, G: ***C*** → **D** then a natural transformation h: F ⇒ G is a |**C**|-indexed family if morphisms h: <h_A: A |F| → A |G| >$_{A \in |C|}$ such that for all a ∈ **C** (A, B), we have aFh_B = h_A(aG), or that the arrow diagram commutes in **C.**

A |F| →...h_A → A |G|

|
αF
|

|
αG
|

B |F| → ...h_B → B |G|

Newer areas to consider from the above, we can conclude, for example, that for any theory **T**, if A: $\mathbf{T}^{op}$ → **Set** is a **T**-algebra, then there is exactly one **T**-homomorphism h: **T**(_,0) ⇒ A, where **T**(_,0) is the initial **T**-algebra.

5.6 INITIAL ALGEBRAIC THEORIES AND COMPUTABLE TREES

In this section, we present methods of constructing initial models by algebraic tree rewriting from the author's decade before. Thus we show how initial algebras can be defined by subtree replacement and tree rewriting. Subtree replacement systems in computing and for equational theories and theorem proves have been for quite some time a subject of research by this author and many others c.f. references. We present a brief overview and formulation of algebraic subtree replacement systems and connect it to our recent computational views with generic diagrams (Boone, 1959; Nourani, 1985, 1991).

We formulate subtree replacement systems with algebraic theorems [ADJ 1975] in the present chapter. First we define algebraic subtree replacement. Start with a set A of Σ-terms for an S-sorted signature Σ. Let E be a set of Σ-equations. Let S* denote the set of all strings u = u1…un. For ui in S, including the empty string; let S+ denote the set of all nonempty strings of sorts. For si n S and u=u1…un in S*, let A(u, s) denote the set of all Σ-expressions of sort s, with variables (usually x1, …, xn) of sorts u = u1…un. For u in S* and v in S+, with v = v1…vm, let A(u, v) denote the set of all tuples e = (e1, …, em) with ei in A(u, vi) Σ v-tuple expression in u-tuple variables; wrote e:u → v.

Then A is the union over all s in S of A s = A (l, s), where l denotes the empty string.

Define a set of Σ-rewrite rules (or Σ-equations) for an algebra A of signature Σ, to be a set of pairs (f, g) with f, g in some A(u, s), u in S* and s in S. Let E(u, s) be the set of all rules (f, g) with (f, g) in A(u, s). To define the set R[E] of applications of rules in E to constant expressions, i.e., expressions without variables, we need to define the notion of substitution.

Given f: u → v and h: w → u, let foh denote the result of substituting the ui for the corresponding variable xi in f, yielding f ° h: w → v. Then R[E] is the set of all pairs (q ° f ° h, q ° g ° h) such that (f, g) in E(u, s), q in A(s, s'), and h in A(l, u). This is all pairs resulting from substituting a tuple h of constant expressions (without variables) into (f, g), then substituting the result for a variable in some expression one instance at a time. We refer to (A, R[E]) as an algebraic subtree replacement system, and say that q o g oh results from rewriting q o f oh.

We say that <q, h> is a substitution. This formulation is following the usual practice in algebraic theories of defining substitution as tree composition. As an example, let f = x1+(x1+x2)+x3. Then f =g is the associative law. An application of this rule rewrites (2+((3+5)+1)*8) to ((2+(3+5))+1)*8. Here q = x1*8 and h = (2, (3+5),1). The best way to visualize this process is to look at the tree representation of the above expressions.

Given an algebraic subtree replacement system (A, R[E]), the relation defined by R[E] on A will be denoted by "=>." We write t => t', for t and t' in A, iff t can be rewritten to t' by a finite number of applications of the replacement rules in (A, R[E]). Let => * denote the reflexive transitive closure of =>. Then we say that (A, R[E]) is *Church-Rosser* iff for all *a*, *b*, *c* in A such that *a* =>**b* and *a* =>**c*, there is *a, d* in A such that *b* => **d* and *c* =>**d*.

An algebraic replacement system has the finite termination property : FTP iff there is no infinite sequence of rewrites x1 => x2 => … A term t in A is said to be in normal form with respect to (A, R[E]) if it cannot be rewritten by the rules of (A, R[E]). Such term t is said to be *irreducible.* If reduced forms are unique, we say that (A, R[E]) has the *unique termination property* (UTP). The significance of the UTP and FTP is that under such circumstances unique normal forms exist.

Definition 5.17 Let Σ be an S-sorted signature, E a set of Σ-equations. An *equational theory* with signature Σ and axioms E is a formal system with Σ-equations as its only formulas, with axioms that are those in E, and rules of inference that are reflexivity, symmetry, transitivity, and substitution properties of equality, i.e.,

1. t =t 2. t =t' => t'=t 3. t =t' and t'=t' => t=t" 4. t =t' => g ° t ° h = g ° t' ° h, where <q, h> is a substitution.

Here t, t', and t' are Σ -terms (including variables).

Definition 5.18 Let T be an equational theory with signature Σ and axioms E, with Σ an S-sorted signature. A proof of t=t' in T is a finite sequence *b* of Σ ending in t=t' such that if q=q' is in *b*, then either q=q' in E, or q=q' is derived from 0 or more previous equations in E by one application of the rules of inference. Write T <ST>|—t=t' for "T proves t=t' by algebraic subtree replacement system."

When the set of axioms E is not specific, and is only an implicit set, we write R instead of R[E]. Deduction with equational theories is closely related to rewrite rule simplifications. Given a set R of Σ-equations, for a signature Σ, let Ro be the union of R with its converse. Furthermore, let R be Ro viewed as a set of rewrite rules, that is, l =r in Ro is viewed as l => r in R. Let $T^{\rightarrow}$ be the equational theory of signature Σ and axioms R. The following lemma expresses a relationship between $T^{\rightarrow}$ and R that clarifies the theoretical development and is useful for our applications.

Lemma 5.1 Let R be a set of Σ-equations. Let R be the set of algebraic Σ-rewrite rules obtained by considering each equation l =r in Ro as a rule l => r, then for t, t' in T<Σ>, t => * t' iff $T^{\rightarrow}$ <ST>|–t = t'.

Proof (Exercise).

Recall that a presentation (Σ, E) defined an equational theory of signature Σ and axioms E. Next we show how canonical models can be constructed by algebraic subtree replacement system. A definition and what we have done thus far gets us to where we want to go: the canonical algebraic term rewriting theorems from over a decade ago, for example, (Nournai, 1985, 1996). Σ<s1, …, sn, s> denotes the part of the signature with operations of arity (s1, …, sn) and coarity s, with Csi the carrier of algebra C of sort si.

Definition 5.19 Let R be a convergent set of Σ-rewrite rules, for example, T <Σ, R> has the finite termination property- FTP and the unique termination property-UTP, let [t] denote the R-reduced form of t in T<Σ>. Let |C| be a subset of |T<Σ>|, for g in Σ <s1…sn, s> and ti in C si, define g_C (t1, …, tn) = [g(t1, …, tn)]. If this always lies in C, then C becomes a Σ-algebra, and we say that (C, R) represents a Σ-algebra A iff the Σ-algebra so defined by (C, R) is Σ-isomorphic to A.

The following intermediate theorem gives *sufficient conditions for constructability* of an initial model [Nourani, 78,79, 82b] for an equational presentation. It is the mathematical justification for the proposition that initial models can be automatically implemented (constructed) algebraic subtree replacement systems with normal forms defined by a minimal set of functions that are Skolem functions or type constructors.

Theorem 5.7 Let Σ be an S-sorted signature, R a convergent set of Σ -rewrite rules. Let |C| be a subset of |T<Σ>|. Define gC(t1, …, tn) = [g(t1, …, tn)]. Furthermore, assume that [f] = f for all f in Σ(l, s) If there exists a subset CF of Σ such that |C| = |T<CF>| and the following conditions are satisfied:

1. g_C(t1, …, tn) in C whenever ti in C, where g has nontrivial arity (s1, …, sn), for ti of sort si;
2. for all g of nontrivial arity (s1,.., sn) with ti in C, g in CF, g_C(t1, …, tn) = gt1, …tn; in particular for a constant g, g_C = g;
3. for g in Σ – CF of arity (s1,.., sn), gC(t1, …, tn)=t, for some t in T<CF>; Then: (i) C is a canonical term algebra; and (ii) (C, R) represents T <Σ, R>, R is R viewed as a set of Σ-equations. Proof (Nourani, 79).

Theorem 5.8 Let Σ be an S-sorted signature, and R a convergent set of rewrite rules such that [g] = g. Define a Σ-algebra structure C on T<Σ> by g_C (t1, …, tn) = [g(t1, …, tn)]. Let C* be the smallest sub Σ-algebra of C. Then C is a canonical term algebra consisting of R normal forms and (C, R) represents T <Σ, R>.

Proof (author, e.g., 1996)

Corollary 5.4 Let Σ be an S-sorted signature, R a convergent set of Σ-rewrite rules then. Let |C| be subset of |T<Σ>|. Define gC (t1, …, tn) = [g(t1, ….tn)]. If there exists a subset CF of Σ such that |C| is subset of |T<Σ>|, and

1. g_C(t1, …, tn) in C whenever ti in C, for g of nontrivial arity (s1, …, sn);
2. for all g in Σ of nontrivial arity (s1, …, sn) with ti in C, g_C (t1, …, tn)= gt1…tn; in particular g_C = g for g a constant symbol;
3. for g in Σ – CF, g_C(t1, …, tn) = t for t in C; then: (i) C is a canonical term algebra; and (ii) C with R represents T <Σ, R> Proof (Nourani79).

Note that CF does not appear in the statement of the above theorem, but it is implicit in claiming a C*. CF is how we could get a C*.

The above theorems and their corollary point out the importance of the constructor signatures in computational characterization of initial models. These are the minimal set of functions that by forming a monomorphic pair with the base set, bring forth an initial model by forming the free trees that define it.

Thus an *initial free model* is formed. The model than can be obtained by algebraic subtree replacement systems (Klop, 2010; Nourani, 1985, 1996; Terese, 2003).

The generic diagram for the model is also defined from the same free trees. The conditions of the theorems are what you expect them to be: that canonical subset be closed under constructor operations, and that operations outside the constructor signature on canonical terms yield canonical terms. The group example, cyclic group of order six on the set {*x*, *y*}, is revisited here again: The normal forms are the six element of the set {e, *x*, *y*, *xy*, *y*2, *xy*2}. Now consider the following three basic group axioms:

$e.x = x\ x{-}1.x = e\ (x.y).\ z = x.\ (y.z)$

Its completion thorough the Knuth-Bendix algorithm (Knuth Bendix, 1969) is attained by the following additional equations from (Nourani, 1993a).

e–1 = e (y–1)–1 = y (y.y")–1 = y"1.y y–1.(y.z) = z z.e= z y. (y–1.z) = z y.y–1 = e

Using the above equations as a set of left-right rewrite rules R, the group, the group structure on $\{x, y\}^*$ with R defines a canonical group structure C isomorphic to the cyclic group of order six. The set of type constructors for the given group along with the set R or rewrite rules satisfies the conditions of Theorem 3.5. Thus

(C, R) is isomorphic to the cyclic group structure and represents the initial tree algebra T<Σ, R>.

5.7 EXERCISES

1. The concrete free theory is a category identifying each *a in Ω n. What are the morphisms.*
2. If N[Ω] is the concrete free theory with base W and K is a theory congruence on N[Ω] then the factor theory N[Ω]/K is a category.
3. Prove theorem 5.1.
4. The set of all theory congruencies on an algebraic theory **T** forms a complete lattice with respect to the ordering <=.
   Hint: Take an intersection on the w indexed theory congruencies on T: $K_{n,m,}$ n, m ∈ w and show that defines a theory congruence, for the family <K^i| i ∈ I>.
5. Show that the theory of dense linear orderings without end points admits elimination of quantifiers. (Do this by induction on the number of quantifiers, working from the inside out.) Once this has been done, completeness follows easily.
   Hint: axioms for linear orderings: $\forall x \forall y \forall z(x < y \land y < z \rightarrow x < z)$ $\forall x \forall y(x < y \rightarrow \neg y < x)$ $\forall x \forall y(x < y \lor x = y \lor y < x)$
6. Prove the Los theorem on ultrafilters.
   Hint: Induction on complexity of terms.

7. Prove that an ultraproduct of ultraproducts is isomorphic to a single ultraproduct.
8. Prove that every ultrapower embedding is an elementary embedding, every ultrapower extension is an elementary extension, and every ultrapower of A is isomorphic to an ultrapower extension of A.
9. Let T be the theory of groups, then **T^** is isomorphic in (**Cat**) to the category of groups.
   Hint: Recall **T** is isomorphic **to N[Ω]/K** where W = {e}, $\{^{-1}\}$, {.}, Æ, …, Æ and K is the smallest congruence on **N[Ω].**
10. Generalizing from Section 5.7.2, prove that for all a ∈ ***T***(n, m) $(\alpha^{op} A)\, h_n = h_m (\alpha^{op} B)$.
11. Prove Theorem 6.11 (Horn Consistency) Let L1, L2 be two Horn languages. Let L = L1 ∩ L2. Suppose T is a complete theory in L and T1 ⊃ T, T2 ⊃ T are consistent in L1, L2, respectively. Then T1 ∪ T2 has a model M that is definable from a generic (finite) direct product diagram on the language L1 ∪ L2.
    Hint: Apply Robinson consistency with direct product Horn filters.
12. Prove lemma 3.1
    Hint: Apply Robinson consistency with direct product Horn filters.
12. Prove lemma 3.1.

KEYWORDS

- **functorial algebraic theory**
- **isomorphic structures**
- **ultrafilter**
- **ultraproduct construction**

CHAPTER 6

GENERIC FUNCTORIAL MODELS AND TOPOS

CONTENTS

Topos, plural "topoi" or "toposes", is a type of category that behaves like the category of sheaves of sets on a topological space (or more generally: on a site). A typical example is the **Grothendieck topoi** that have applications in algebraic geometry and sets. Since the introduction of sheaves into mathematics in the 1940s a major theme has been to study a space by studying sheaves on a space. This idea was carried onto what is called Grothendieck topos. The main utility of this notion is in the abundance of situations in mathematics where topological intuition is very effective but a natural topological space is lacking. The single success of this programmatic idea has been the introduction of the étale topos of a scheme.

6.1 ELEMENTARY TOPOI

The axiomatic foundation of mathematics is set theory, in which all mathematical objects are ultimately represented by sets (even functions which map between sets). More recent work in category theory allows this foundation to be generalized using topoi; each topos completely defines its own mathematical framework. The category of sets forms a familiar topos, and working within this topos is equivalent to using traditional set theoretic mathematics. But there are alternative topoi. A standard formulation of the axiom of choice can be stated in any topos, and there are topoi in which it is invalid.

It is also possible to encode an algebraic theory, such as the theory of groups, as a topos, in the form of a classifying topos. The individual models of the theory, that is, the groups in our example, then correspond to functors from the encoding topos to the category of sets that respect the topos structure.

When used for foundational work a topos will be defined axiomatically; set theory is then treated as a special case of topos theory. Building from category theory, there are multiple equivalent definitions of a topos. The following has the virtue of being concise:

A topos is a category, which has the following two properties:

1. All limits taken over finite index categories exist.
2. Every object has a power object. This plays the role of the powerset in set theory.

Formally, a *power object* of an object X is a pair (PX, $\backslash_X$) with $\backslash_X \subseteq$ PX X X, which classifies relations, in the following sense. First note that for every object I, a morphism h: I → PX ("a family of subsets") induces a subobject {(I, *x*) | *x* ∈ h(i)} Í I *x* X. Formally, this is defined by pulling back $\backslash_X$ along r *x* X: I *x* X → PX *x* X. The universal property of a power object is that every relation arises in this way, giving a bijective correspondence between relations R ∈ I *x* X and morphisms h: h: I → PX.

From finite limits and power objects one can derive that:

- all colimits taken over finite index categories exist.
- any two objects have an exponential object.
- the category is Cartesian closed.

In some applications, the role of the subobject classifier is pivotal, whereas power objects are not. Thus some definitions reverse the roles of what is defined and what is derived.

A topos as defined above can be perceived as a Cartesian closed category for which the notion of subobject of an object has an elementary or first-order definition. This notion, as a natural categorical abstraction of the notions of subset of a set, subgroup of a group, and more generally subalgebra of any algebraic structure, was realized before topos. That is definable in any category, not only topoi, in second-order language, that is, in terms of classes of morphisms. Given two monics *m*, *n* from respectively *Y* and *Z* to *X*, write $m \leq n$ when there exists a morphism p: $Y \rightarrow Z$ for which $np = m$, inducing a preorder on monics to *X*. When $m \leq n$ and $n \leq m$, *m* and *n* are considered equivalent. The subobjects of *X* are the resulting equivalence classes of the monics to it.

In a topos "subobject" becomes, implicitly, a first-order notion, a topos is a category *C* having all finite limits and hence in particular the empty limit or final object 1. It is then natural to treat morphisms of the form *x*: 1 → *X* as *elements* $x \in X$. Morphisms f: $X \rightarrow Y$ thus correspond to functions mapping each element $x \in X$ to the element $fx \in Y$, with application realized by composition.

One might then think to define a subobject of *X* as an equivalence class of monics m: $X' \rightarrow X$ having the same image or range $\{ mx \mid x \in X' \}$. However, two or more morphisms can correspond to the same function. One cannot assume that *C* is concrete in the sense that the functor *C*(1,-): *C* → **Set** is faithful. For example the category **Grph** of graphs and their associated homomorphisms is a topos whose final object 1 is the graph with one vertex and one edge (a self-loop), but is not concrete because the elements $1 \rightarrow G$ of a graph G correspond only to the self-loops and not the other edges, nor the vertices without self-loops. The second-order definition makes G and its set of self-loops with distinct subobjects of G. Topoi provide a more abstract, general, and first-order solution.

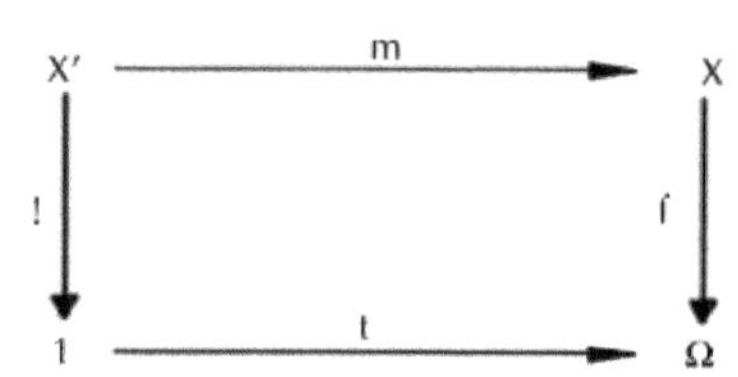

FIGURE 1 *m* as a pullback of the generic subobject *t* along *f*.

The topos C has a subobject classifier Ω, namely an object of C with an element $t \in \Omega$, the *generic subobject* of C, having the property that every monic $m: X' \rightarrow X$ arises as a pullback of the generic subobject along a unique morphism $f: X \rightarrow \Omega$, The pullback of a monic is a monic, and all elements including t are monics since there is only one morphism to 1 from any given object, whence the pullback of t along $f: X \rightarrow \Omega$ is a monic. The monics to X are therefore in bijection with the pullbacks of t along morphisms from X to Ω. The latter morphisms partition the monics into equivalence classes each determined by a morphism $f: X \rightarrow \Omega$, the characteristic morphism of that class, which we take to be the subobject of X characterized or named by f.

In the concrete case, with $C(1, -)$ faithful, like the category of sets, the situation behaves like functions. Here the monics $m: X' \rightarrow X$ are exactly the injections (one-one functions) from X' to X, and those with a given image $\{ mx \mid x \in X' \}$ constitute the subobject of X corresponding to the morphism $f: X \rightarrow \Omega$ for which $f^{-1}(t)$ is that image.

To summarize, this first-order notion of subobject classifier implicitly defines for a topos the same equivalence relation on monics to X as had previously been defined explicitly by the second-order notion of subobject for any category. The notion of equivalence relation on a class of morphisms is itself intrinsically second-order, which the definition of topos neatly sidesteps by explicitly defining only the notion of subobject *classifier* Ω, leaving the notion of subobject of X as an implicit consequence characterized (and hence namable) by its associated morphism $f: X \to \Omega$.

Giraud theorem is an equivalence statement. The following are equivalent:

There is a small category D and an inclusion C A Presheaf(D) that admits a finite-limit-preserving left adjoint. Presh(D) denotes the category of contravariant functors from D to the category of sets; such a contravariant functor is frequently called a presheaf.

C is the category of sheaves on a Grothendieck site. C satisfies Giraud's axioms, below.

A category with these properties is called a "(Grothendieck) topos". Here if X and Y are topoi, a *geometric morphism* $u: X \to Y$ is a pair of adjoint functors (u^*, u_*) (where $u^*: Y \to X$ is left adjoint to $u_*: X \to Y$) such that u^* preserves finite limits. Note that u^* automatically preserves colimits by virtue of having a right adjoint.

By Freyd's adjoint functor theorem, to give a geometric morphism $X \to Y$ is to give a functor $u^*: Y \to X$ that preserves finite limits and all small colimits. Thus geometric morphisms between topoi may be seen as analogs of maps of locales.

If X and Y are topological spaces and u is a continuous map between them, then the pullback and push forward operations on sheaves yield a geometric morphism between the associated topoi. A point of a topos X is defined as a geometric morphism from the topos of sets to X.

If X is an ordinary space and x is a point of X, then the functor that takes a sheaf F to its stalk F_x has a right adjoint (the "skyscraper sheaf" functor), so an ordinary point of X also determines a topos-theoretic point. These may be constructed as the pullback-pushforward along the continuous map $x: 1 \to X$.

A geometric morphism (u^*, u_*) is *essential* if u^* has a further left adjoint $u_!$, or equivalently (by the adjoint functor theorem) if u^* preserves not only finite but all small limits. Another important class of ringed topoi, besides ringed spaces, are the etale topoi of Deligne-Mumford stacks.

6.2 GENERIC FUNCTORIAL MODELS

The functor F: $\boldsymbol{L}_{\omega 1, K} \rightarrow$ *Set* had been defined since 1994 by a list of sets and functions on strings on the category defined on the infinitary language $L_{\omega 1, \omega}$. Let us refer to the above functor by the *generic model functor*. Restating a couple definition for self-containment here towards defining D<A, G> models categories.

Definition 6.1 Let M be a structure for a language L, call a subset X of M a generating set for M if no proper substructure of M contains X, that is, if M is the closure of X U {*c*(M): *c* is a constant symbol of L}. An assignment of constants to M is a pair <A, G>, where A is an infinite set of constant symbols in L and G: A → M such that {G(a): a in A} is a set of generators for M. Interpreting a by g(a), every element of M is denoted by at least one closed term of L(A). For a fixed assignment <A, G> of constants to M, the diagram of M, D<A, G>(M) is the set of basic (atomic and negated atomic) sentences of L(A) true in M.

Definition 6.2 A generic diagram for a structure M is a diagram D<A, G>, such that the G in definition above has a proper definition by a specific function set, that is, Σ_1 Skolem functions. –

To save from confusion we have referred to category arrow diagrams by A-diagrams, to distinguish from model diagrams, since definition 6.1. Let *D<A, G>* be the category of models definable with generalized diagram D<A, G> and their morphisms. Recall from preceding chapters that K<A> is the least fragment that contains L<A>. Each formula in K<A> contains only finitely many *c* in C. We refer to the fragment by $L_{\omega 1, K.}$ Carrying on with the small-complete category $\boldsymbol{L}_{\omega 1, K}$ from $L_{\omega 1, K}$ The category is the preorder category defined by the formula ordering defining the language fragment. The objects are small set fragments. There are three categories at play- the category $\boldsymbol{L}_{\omega 1, K,}$ the category **Set,** and the category **D<A, G>.**

The **D<A, G>** category is the category for models definable with D<A, G> and their morphisms. The techniques we are presenting by the three categories save us from having to yet develop a categorical interpretation for $\mathbf{L}_{\omega 1, K}$, in categorical logic as in (Lawvere's, 1967) Chapter 5, and categorical logic, c.f. Chapter 1. A computing view to the functorial models are presented by defining Hasse diagrams on the $\boldsymbol{L}_{\omega 1, K}$ fragments. The limit model is defined by computing Hasse diagram limits and natural transformations on the limit cones.

Objects: $\boldsymbol{L}_{\omega 1, K}$ Models definable with D<A, G>;

Morphisms: H: <h: M → M'>, where h is a model homomorphism.

An embedding from a model definable in *D<A, G>* to the limit created from the functor from the Infinitary language category $\boldsymbol{L}_{\omega 1, K}$ to Set can be defined.

Proposition 6.1 There is a homomorphic embedding from every model M in category D<A, G> to the generic functor model on $\boldsymbol{L}_{\omega 1, K} \rightarrow$ ***Set.***

Proof Exercise.

Now let us define a functor V: $\boldsymbol{L}^{Op}_{\omega 1, K} \rightarrow$ *Set* to be the generic functor defined above. Define a functor F: *D<A, G>* → $\boldsymbol{L}_{\omega 1, K}$ by universal embedding from the diagram functions. We prove that V creates a limit for F. A generic model is defined by the created limit. The functor from $\boldsymbol{L}_{\omega 1, K}$ to Set is an example generic functor. The functor was proved to have limiting models that are positive generic models in the preceding chapters. Positive forcing is applied to further fill the gap between forcing and toposes that had started to be filled with apparent since our Functorial Model theory project. Models are generated with lifting model diagrams to categories, called DAG categories. Models are generated at limit cones on D<A, G>.

6.3 GENERIC FUNCTORS

Recall the small-complete category $\mathbf{L}_{\omega 1, \mathbf{K}}$. The category is the preorder category defined by the formula ordering defining the language fragment. The objects are the small fragment sets, and the arrows the preorder arrows.

The functor F: $\mathbf{L}^{OP}_{\omega 1, \mathbf{K}} \rightarrow$ Set had been defined by a list of sets and functions on strings on the infinite language category $\mathbf{L}_{\omega 1, \mathbf{K}}$.

Lemma 6.1 $L_{\omega 1, K}$ is a small-complete category.

The lemma was proved in Chapter 3, but the topos properties are interesting to glimpse on here. The proof applies the sort of techniques applied by the present author to define generic sets with the fragment $L_{\omega 1, K}$. To prove $\mathbf{L}_{\omega 1, \mathbf{K}}$ is small complete, that is, every small diagram has a limit, we can defined a functor.

H: $_{\omega}$ Op$\rightarrow\mathbf{L}_{\omega 1, \mathbf{K}}$ and created a limit in $\mathbf{L}_{\omega 1, \mathbf{K}}$ as follows. The functor H is a list of sets Hn, consisting of the strings in $L_{\omega 1, K}$, and functions fi:Hi+1 → Hi. Form the product set iHi. Let l = {l0, l1, l2, ...}, with each ln ∈ Hn, and the projections pn: Hi → Hn, forming the following diagram.

To form a limit at a vertex for commuting with the projections take the subset S of the strings in $L_{\omega 1, K}$, which match under f, that is, f sn+1 = sn, for all n. The above A-diagram is well- defined because sets which are generic with respect to a base set, for example the intended limit S are definable, and we only have to present limits for small A-diagrams.

The completeness proof, that is, that every functor from a small category to $L_{\omega 1, K}$ has a limit, has bearings to the completeness proof for the category Set. The crux of the proof is there being limiting cones for arbitrary functors F: J → $L_{\omega 1, K}$ for the K<A> fragment. We can define cones to base F from {}. The natural transformations are on functors sending arbitrary objects to sets on fragment string sets, arrowed by preorder functions. Since sets that are generic with respect to the base set are always definable by what the author had put forth in (Nourani 1983, Nourani 1997), (also see Theorem 4.3, Section 4,) the limiting cones can always be defined. Cones to base F from {} pick fragment sets. A limiting cone can always be defined with natural transformations on the preorder functors.

Define a functor F: $L_{\omega 1, K}$ → Set by a list of sets Mn and functions fn. The sets correspond to an initial structure on $L_{\omega 1, K}$. For example, to f(t1, t2,.. tn) in $L_{\omega 1, K}$ there corresponds the equality relation $f(t_1, t_2, \ldots, t_n) = ft_1 \ldots t_n$. Let us refer to the above functor by the generic model functor since we can show it can defines generic sets from language strings to form limits and models. It suffices to define the functor up to an initial model without being specific as to what the model is, to have a generic functor.

Theorem 6.1 The generic model functor has a limit.

Proof Chapter 3.

We referred to the above functor by the generic model functor since it defines generic sets from language strings to form limits and models. The proof for that theorem follows from the sort of techniques applied by the present author to define generic sets with the fragment $L_{\omega 1, K}$ by this author 1982, 1983.

The model theoretic properties are scattered on chapters 3–5 and further explored on toposes here.

6.4 INITIAL MODELS

Lemma 6.2 The model defined from the G-diagram is initial in D<A, G>.

Proof Follows from section 6.2 and the relation between canonical models generic sets and initial models via definition and since a canonical initial model in the set theory sense is defined by the G-diagram.

Let D(A, G) be the category of models definable with generalized diagram D<A, G> and their morphisms. An embedding from a model definable in D(A, G) to the limit created from the functor from the infinitary language category $L_{\omega 1, K}$ is defined. Define a functor F: D(A, G) → $L_{\omega 1, K}$ by universal embedding from the diagram functions.

A computing view to the functorial models are presented by defining Hasse diagrams on the $L_{\omega 1, K}$ fragments. The limit model is defined by computing Hasse diagram limits and natural transformations on the limit cones.

Lemma 6.3 The model defined from the generic diagram is initial in D(A, G).

Define a functor F: $L_{\omega 1, K}$ →Set to be the generic functor.

Define a functor V: D(A, G) → $L_{\omega 1, K}$ by universal embedding from the diagram functions.

As an example a theorem might be proved an alternate way to define functorial models.

Definition 6.3 A functor V:A → X creates limits for a functor F: J → A if every limiting cone t: x → VF in X there is exactly one pair <a, s > consisting of an object a in A with Va=x and a cone s: a –> F with Vs = t and if moreover this cone s:a →F is a limiting cone in A.

Theorem 6.2 V creates a limit for F.

Proof the functor **V**: D(A, G) →$L_{\omega 1, K}$ is on the category of models definable in D(A, G). The models generated are similar to generic set foundations models.

To prove it we have to review some theorems we had proved in the author's earlier papers, that is, Chapter 3, on definitions and theorems put forth topological properties for the infinite language category $L_{\omega 1, K}$, to create limits with generic sets. Á-genericity is a topological property defined on an infinitary language, $L_{\omega 1, K}$.

Next, we review what the functor V: D<A, G> → $L_{\omega 1, K}$. The functor V is from a model definable in D<A, G> – the category of models definable with generalized diagram D<A, G> to the limit created from the functor from the infinitary language category $L_{\omega 1, K}$ to Set. D<A, G> is similar to the diagrams defining generic models for set theory. The sets affected are the set of formulas defined by the definition of the fragment $L_{\omega 1, K}$.

Define a functor F: $L_{\omega 1, K}$ → Set to be the generic functor defined in this chapter before.

Define a functor V: D<A, G> → $\mathbf{L}_{\omega 1, \mathbf{K}}$ by universal embedding from the diagram functions.

Creating limits amounts to defining generic sets on $L_{\omega 1, \omega}$ for $L_{\omega 1, K}$. The present theory creates limits for Infinitary languages, presenting limit topology for the language categories. Generic sets for infinitary languages have been applied by this author to define Positive Forcing. Positive forcing is not applied by the direct here. However, the language sets from which the limits are definable have been shown to be dense by this author thus pointing us to the infinitary language topology, e.g. Chapter 3. To define language topology we present the definitions for positive forcing and positive genericity that we had defined since 1981. The applications to categories and toposes are ever since 1994 in the author's publications.

Lemma 6.4 The model defined from the G-diagram, d<F, M> is initial in D<A, G>.

VF is defined by $L_{\omega 1, K} \rightarrow$ D<A, G>.

The Theorem's proof revisited:

V creates limits means V defines limits for functors F whose composition VF already has a limit.

VF is defined by $L_{\omega 1, K}$ is an S-indexed family of arrows with vertex in VF and a base set s in $\mathbf{L}_{\omega 1, \mathbf{K}}{}^{op} \rightarrow$ S. The cone s: a –> F is a limiting cone in $L_{\omega 1, K}$. Since a generic diagram instance is defined by a unique function set, a generic model is defined by the created limit.

6.5 POSITIVE FORCING MODELS

The forcing property defined for positive forcing is similar to what Keisler's model-theoretic forcing is defined on algebraic structure.

Definition 6.4 A forcing property for a language L is a triple I = $(S,\leq,f)$ such that (i) $(S,\leq)$ is a partially ordered structure with a least element, for example, 0; (ii) f is a function which associates with each p in S a set f(p) of atomic sentences of L[C]; (iii) whenever $p \leq q$, $f(p) \subseteq f(q)$; (iv) let I and t be terms of L[C] without free variables and p in S, φ a formula of L[C] with one free variable. Then if (I$\neg$=t) is in f(p) then (t=I) is in f(q) for some $q > p$. φ(t) in f(p) implies φ (l) is in f(q) for some $q \geq p$. For some c in C and $q > p$, (c=l) is in f(q). The elements of S are called conditions for F.

Definition 6.5 The relation p ||-+ φ read "p positively forces φ" is defined for conditions p an q in S as follows: for an atomic sentence φ, p ||-+ φ iff φ is in p; for an open formula φ of the form f(X) = g(X), p ||-+ φ, iff for all c in L[C], where c is an n-tuple of constants form L[C] and all q, $p < q$ implies there is an r such that $r > q$ and r ||-+ f(c) = g(c).

Definition 6.6 Let I = (S,≤,f) be a positive forcing property, we say that a subset G of S is positively generic iff

(i) p in G and q < p implies q ∈ G; (ii) p,q∈ G, implies there is an r in G such that p ≤ r and q ≤ r (ii) for each sentence φ there exists p in G, such that either p ||-+ φ or there is no q in S, q >p such that q ||-+ φ.

A special case of the above definition is when S consists of sets of formulas. For such S , we can make a substitution: subset relation for <. Clause (iii) can then be stated:

(iii) for each sentence φ, there exists p in G such that either p ||-+ φ or p U {φ} is not a condition for F.

Definition 6.7 Let Q be a poset, F a family of sets, G subset of Q. the G is F-generic iff the following holds;

(i) whenever p in G and q < p, then q is in G. (ii) whenever p,q in G, then there exists an r in G with r > p and r > q. In particular, any two elements of G are compatible. (iii) G∩D¬=∅ for any dense D subset of Q with D in F.

Note that a subset D of Q is said to be dense iff for every p in Q there is a q in D with p < q. []

Definition 6.8 Let (M, a)*c* in C be defined such that M is a model for L and each constant *c* in C has the interpretation a in M. the mapping *c* → ac is an assignment of C in M. We say that (M, a)*c* in C is canonical model for a presentation P iff the assignment *c* → a maps C onto M, i.e., M=(a:*c* in C).

Definition 6.9 A generic set is said to generate M iff M is a canonical model and every sentence j of L[C], which is forced by some p in G holds in M.

Definition 6.10 M is a generic model for a condition p iff M is generated by some generic set G that contains p.

Definition 6.11 A theory is said to be inductive if T is equivalent to a set of ∀∃ sentences- sentences of the form ∀x1...∀xn∃ x1..∃ xn y, where y is quantifier-free.

Definition 6.12 A G-diagram for a structure M is a diagram D<A, G>, such that the G in definition above has a proper definition by a known set of function symbols.

The following theorem presented itself in the course of writing the proofs for the present paper and is stated. It is a consequence of the results proved by this author in Ref. [3].

Theorem 6.3 The power set of an inductive theory defined on $L_{\omega 1, K}$ generates a generic model for the theory, provided the inductive theory has a defined generic diagram for an initial model for the theory.

Proof If the inductive theory has a definable G-diagram for its initial model M, then its powerset has the inductive closure as a maximal member. Since M is a canonical model, for every sentence forced by some p in the powerset, if j is atomic, then by definition it must be in p to be forced. Thus it is modeled by M. For an arbitrary formula j, if j is forced by some p, then there is an r >=p such that r forces φ(1), where φ(1) is obtained from j by a Skolemization on generic diagram terms for the existential quantifiers. Therefore, M models j and M is a generic model.

The above definitions and theorems put forth topological properties for the infinite language category $L_{\omega 1, K}$, to create limits with generic sets. $\mathfrak{I}$-genericity is a topological property defined on an infinitary language, $L_{\omega 1, K}$.

6.6 FUNCTORS COMPUTING HASSE DIAGRAM MODELS

The fragment sets on $L_{\omega 1, K}$ on which the functor F defines models are viewed by Hasse diagrams defining orderings on fragments. The limits on the Hasse diagrams are specific functorial models generated.

Let us examine what the functors F and VF are. Start with the completeness proof for the fragment category $L_{\omega 1, K}$ to view its limiting.

cones in $L^{OP}_{\omega 1, K}$. The cones to base F are picking sets from $L_{\omega 1, K}$ and the limiting cone is defined by generic sets on $L_{\omega 1, K}$.

For a functor F: J → $\mathbf{L}_{\omega 1, \mathbf{K}}$, j → k corresponds to a preorder arrow on $L_{\omega 1, K}$ example fragments depicted we can define preorders on the fragment sets as follows.

j⟶k ⇒ ∧__/ ⟶ ∧/__

F1F2.....Fn... ⟶

\ | / corresponds to

Fu σj = σk

FuFj ⟵ Fk

↓ σkσj

{}

The limiting cone v: vj: Cone({}, F) → Fj, is defined by σ → σj where a cone to base F vj: {F1 ←F2 ←̃ →Fj

\|/

{}

Corresponding Cone (X, F) to $L_{\omega 1, K}$ (X, Cone({}, F)). From the above we have a view as to how limiting cones can be defined to pick a specific limit model on $L_{\omega 1, K}$. A specific theorem follows as stated.

6.7 FRAGMENT CONSISTENT MODELS

During the 1990s, a functorial model theory has been presented by the author applying positive forcing. An infinitary functorial generalization to Robinson consistency on varying language fragment is applied by the author to present fragment consistent models. Some applications are enclosed from where rudimentary models are presented. Apart from its applications to model theory, forcing companions and Peano arithmetic, positive forcing has since been applied to present a functorial model theory (Nourani 1990 s).. The last decade has since seen many publications on models to Peano arithmetic while the relations to complementary model-theoretic developments on (Knight, 1987, 2001) has not been explored. On the functorial model-theory projects foundational gap between forcing and toposes is filled with a precise progress. Positive forcing and positive tree logic is a practical way to apply forcing to computability. It is defined for tree computing and initial structures. The foundations and the application areas are presented. Positive generic sets and diagrams are further applied via a new functorial model theory to define computable functors and new computability theories with principles parallel to abstract recursion. That area is not treated in this textbook for timing containment reasons.

6.7.1 FILTERS, IFLCS, PRESHEAVES, AND FRAGMENT TOPOLOGY

The author had defined infinite language categories the on the Keisler $L_{\omega 1, K}$ fragments to present a functorial model theory since 1996. Here a basic Grothendeick topology is defined on the fragments and further new categorical areas are presetting. The author's generic functors and functorial model theory are applied to presheaves providing a glimpse onto the functorial models on the topologies.

A site I given by a category C (the underlying category of the site) together with Grothendeick topology on C given by a class Cov (A) for each object A of C.

The elements of Cov(A) are families (sets)$(A_i f_i \rightarrow A)_{i \in I}$ of morphisms with domain A. An element of Cov(A) is called a covering family of A, $A \in C$.

The topology has to satisfy the following conditions: (note the mapping arrows are prefixed with the mapping name, for example $f_i \rightarrow A$)

Every isomorphism A' $f \rightarrow A$ gives an elementary covering family {A' $f \rightarrow$ A}$\in$Cov(A)(Satisfiability under pull-backs), whenever$(A_i f_i \rightarrow A)_{i \in I} \in$ Cov(A) and $Bg \rightarrow A$ is an isomorphism in C then $(A_i x\, Bf_i, B)_{i \in I \in}$ Cov(B) is any pullback diagram for each $I \in$ I.

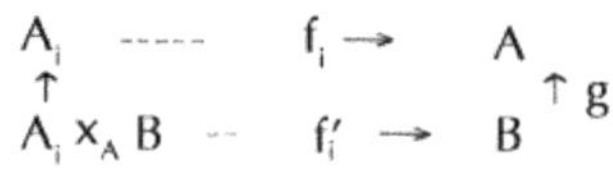

(i) "Closure under composition" whenever $(A_i\ f_i \rightarrow A)_{i \in I}$ Cov(A) and
(ii) $(A_{ij} g_{ij} \rightarrow Ai)_{j \in J\, i} \in$ Cov(Ai) for every $i \in I$? we have that $(A_{ij} f_i {}^\circ g_{ij} \rightarrow A)_{j\, Ji,\, i \in I}$ belongs to Cov(A).
(iii) "Montonicity" If $(A_i f_i \rightarrow A) \in$ Cov(A) and $(B_j g_j \rightarrow A)_{j \in J}$ is such that from any $j \in J$ there is an I and a morphism $B_j \rightarrow A_i$ with

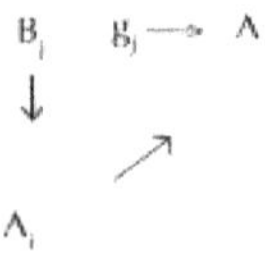

Then $(B_j g_j \rightarrow A) \in$ Cov(A).

Pullback satisfiability on the fragments follows from the preorder morphims- and the embedding lemma that when R and R' be models for L, then R is isomorphically embedded in R' iff R' can be expanded to a model of the model diagram of R. Thus the embedding is in Cov (B).

Composition rule follows from the subfragment definitions follows from the subfragment preorder transitivity.

Theorem 6.4 (Author 2007)

There is a natural Gorothedieck topology on the category $\mathbf{L}_{\omega 1, \mathbf{K}}$.

The objects are fragment sets. Each object A has the Cov(A) are families (sets)$(A_i f_i \rightarrow A)_{i \in I}$ of preorder morphisms on the fragment sets.

The Grotehdieck conditions are checked as follows:

(i) Every isomorphism, is a trivial preorder morphism,A' f→A gives an elementary covering family {A' f A}∈Cov(A).

Pullback satisfiability on the fragments follows from the preorder morphims-and the embedding lemma that when $\mathfrak{R}$ and $\mathfrak{R}$' be models for L, then $\mathfrak{R}$ is isomorphically embedded in $\mathfrak{R}$'iff$\mathfrak{R}$' can be expanded to a model of the model diagram of$\mathfrak{R}$. Thus the embedding is in Cov(B).

Composition rule follows from the subfragment defintions follows from the subfragment preorder transitivity. Every Grothendieck topos is an elementary topos, but the converse is not true (since every Grothendieck topos is cocomplete, which is not required from an elementary topos). The categories of finite sets, of finite *G*-sets (actions of a group *G* on a finite set), and of finite graphs are elementary topoi which are not Grothendieck topoi.

If *C* is a small category, then the functor category **Setcov** *C* (consisting of all covariant functors from *C* to sets, with natural transformations as morphisms) is a topos. For instance, the category **Grph** of graphs of the kind permitting multiple directed edges between two vertices is a topos. A graph consists of two sets, an edge set and a vertex set, and two functions s, *t* between those sets, assigning to every edge *e* its source s(*e*) and target *t*(*e*). **Grph** is thus equivalent to the functor category **Setcov** *C*, where *C* is the category with two objects *E* and V and two morphisms s, *t*: $E \rightarrow$ V giving respectively the source and target of each edge.

The Yoneda Lemma asserts that C^{op} embeds in **Setcov***C* as a full subcategory. In the graph example the embedding represents C^{op} as the subcategory of **Setcov***C* whose two objects are *V'* as the one-vertex no-edge graph and *E'* as the two-vertex one-edge graph (both as functors), and whose two nonidentity morphisms are the two graph homomorphisms from *V'* to *E'* (both as natural transformations). The natural transformations from *V'* to an arbitrary graph (functor) *G* constitute the vertices of *G* while those from *E'* to *G* constitute its edges. Morphism of graphs can be understood as a *pair* of functions, one mapping the vertices and the other the edges, with application still realized as composition but now with multiple sorts of *generalized* elements. This shows that the traditional concept of a concrete category as one whose objects have an underlying set can be generalized to cater for a wider range of topoi by allowing an object to have multiple underlying sets, that is, to be multisorted.

6.7.2 GENERIC FUNCTORS AND SHEAVES

Creating limits amounts to defining generic sets on $L_{\omega 1, K}$ for L_w. The present theory creates limits for infinitary languages to define generic models with language categories. Generic sets for infinitary languages have been applied by this author to define Positive Forcing (Nourani, 1983) and applied to categories and toposes in (Nourani, 1995, 2000). The completeness proof on lemma 1.1, that is, that every functor from a small category to $L_{\omega 1, K}$ has a limit, has bearings to the completeness proof for the category Set. The crux of the proof is there being limiting cones for arbitrary functors F: J → $\mathbf{L}_w$ for the K<A> fragment. We can define cones to base F from {}. The natural transformations are on functors sending arbitrary objects to sets on fragment string sets, arrowed by preorder functions. Since sets, which are generic with respect to the base set are always definable from positive forcing, the limiting cones can always be defined. Cones to base F from {} pick fragment sets. A limiting cone can always be defined with natural transformations on the prodder functor on L^{Op}.

Next, we review what the functor V: D<A, G> → $L_{\omega 1, K}$. The functor V is from a model definable in D<A, G>— the category of models definable with generalized diagram D<A, G> presented by this author in Ref. [4], to the limit created from the functor from the Infinitary language category $L_{\omega 1, K}$ to Set. D<A, G> is similar to the diagrams defining generic models for set theory. The sets affected are the set of formulas defined by the definition of the fragment $L_{\omega 1, K}$.

Recall from Section 6.4 that the model defined from the G-diagram, d<F, M> is initial in D<A, G>.

Now on vertex-based models from (Nourani, 2000) Oxford Ohio and (Nourani, 2000).

Defintion 6.13 A functor V: A → X creates limits for a functor F: J →A if to every limiting cone t: x →VF in X there is exactly one pair <a, s > consisting of an object a in A with Va=x and a cone s: a →F with Vs = tt and if moreover, this cone s:a → F is a limiting cone in A.

Define a functor F: $\mathbf{L}_{\omega 1, \mathbf{K}}$ → Set to be the generic functor defined in the preceding section.

Define a functor V: D<A, G> → $\mathbf{L}_{\omega 1, \mathbf{K}}$ by universal embedding from the diagram functions.

Next, we review what the functor V: D<A, G> → $\mathbf{L}_{\omega 1, \mathbf{K}}$. The functor V is from a model definable in D<A, G> – the category of models definable with generalized diagram D<A, G> presented by this author in (Nourani, 1995) ECCT and AMS, to the limit created from the functor from the Infinitary language category $\mathbf{L}_{\omega 1, \mathbf{K}}$ → Set. D<A, G> is similar to the diagrams defining generic models for set theory. The sets affected are the set of formulas defined by the definition of the fragment $\mathbf{L}_{\omega 1, \mathbf{K}}$.

From section 6.4 we have that the model defined from the G-diagram, d<F, M> is initial in D<A, G> and VF is defined by $\mathbf{L}_{\omega 1, \mathbf{K}}$ → D<A, G>.

Reviewing the preceding section from Sheaves:

V creates limits means V defines limits for functors F whose composition VF already has a limit.

VF is defined by L_{tt} from definition is an S-indexed family of arrows with vertex in VF and a base set s in OP S. The cone s: a→F is a limiting cone in $\mathbf{L}_{\omega 1, \mathbf{K}}$. Since a G-diagram instance is defined by a unique function set, a generic model is defined by the created limit.

VF is defined by $\mathbf{L}_{\omega 1, \mathbf{K}}$ → D <A, G>.

V creates limits means V defines limits for functors F whose composition VF already has a limit.

Theorem 6.5 The functor V: *D <A, G>* → $L_{\omega 1, K}$ defines a generic model for $L_{\omega 1, K}$.

Fi's are distinct sets defined by the functor F. For cones s, { } the preorder arrow.

Let F be the functor F: $\mathbf{L}^{Op}_{\omega 1, \mathbf{K}}$ → Set, be the generic functor.

Define a functor V: **D<A, G>** → $\mathbf{L}_{\omega 1, \mathbf{K}}$ by universal embedding from the diagram functions.

Define a functor V: D<A, G> → $\mathbf{L}_{\mathbf{w}}$ by universal embedding from the diagram functions.

From the above we note that there is a contravariant functors F: $\mathbf{L}^{Op}_{\omega 1, \mathbf{K}}$ → **Set** to form a presheave.

Lemma 6.5 The limiting cones that are the basis for small completeness for $\mathbf{L}_w$ define example representable presheaves for the objects in $\mathbf{L}_{\omega 1, K.}$

This natural transformation: example is a category with objects cones to base F from {}, with morphism what corresponds to the fragment preorders. The new functor G on the Op cateory $\mathbf{L}_w$ is defined based on isomorphisms.

Theorem 6.6 The natural transformations on functors sending arbitrary objects to sets on fragment string sets, arrowed by preorder functions, with cones to base F from **{}**, on the above on the fucntor F: $\mathbf{L}_{\omega 1, K}{}^{op} \rightarrow$ **Set,** are filter creating representable presheaves for the objects in $L_{\omega 1, K}$.

The obvious direction from here is what comma category realizes that above.

6.7.3 FILTERS ON CATEGORIES

In category theory, *filtered categories* generalize the notion of directed set understood as a category (hence called a directed category; while some use directed category as a synonym for a filtered category).

A category J is **filtered** when it is not empty,

- for every two objects j and j' in J there exists an object k and two arrows f: j → k and f': j → k in J;
- for every two parallel arrows u, v: i → j in J, there exists an object k and an arrow w: j → k such that wu = wv.

A diagram is said to be of cardinality k if the morphism set of its domain is of cardinality k. A category J is filtered if and only if there is a cone over any finite diagram d: D → J more generally, for a regular cardinal k, a category J is said to be k-filtered if for every diagram d in J of cardinality smaller than k there is a cone over d.

A **filtered colimit** is a colimit of a functor F: J → C where J is a filtered category. This readily generalizes to k-filtered limits. An **ind-object** in a category C is a presheaf of sets $\mathbf{C}^{op} \rightarrow$ **Set** which is a small filtered colimit of representable presheaves. Ind-objects in a category C form a full subcategory Ind © in the category of functors $\mathbf{C}^{op} \rightarrow$ **Set**. The category Pro © = Ind $(\mathbf{C}^{op})^{op}$ of proobjects in C is the opposite of the category of ind-objects in the opposite category $\mathbf{C}^{op}$.

There is a dual notion of *cofiltered* category. A category J is cofiltered if the opposite category J^{op} is filtered.

A *cofiltered limit* is a limit of a functor F: J → C where J is a cofiltered category.

6.8 HOMOTOPY THEORY OF TOPOS

Artin and Mazur associated to the site underlying a topos a prosimplicial set up to homotopy. Using this inverse system of simplicial sets one may sometimes associate to a homotopy invariant in classical topology an inverse system of invariants in topos theory. The study of the prosimplicial set associated to the etale topos of a scheme is called étale homotopy theory. Let C be the category of CW-complexes with morphisms given by homotopy classes of continuous functions. For each natural number n there is a contravariant functor Hn: C → Ab, which assigns each CW-complex its nth cohomology group (with integer coefficients). Composing this with the forgetful functor we have a contravariant functor from C to Set. Brown's representability theorem in algebraic topology says that this functor is represented by a CW-complex called an Elinberg Maclane Space. Category theory in homotopy type theory allows more freedom in how you treat equality of objects in a category.

A word of caution from this author's perspective chapters is that on the surface one might be tempted to think of the type saturation notions on categoreis as lifts on saturation on types in the sense of preceding chapters. That is not a position we have "depth perception" on now. The obvious definition of a category A has a type of objects, say $A_{0:}$Type and a family of types of morphisms, say:

hom A: $A_0 \times A_0 \rightarrow$Type

For example, we consider 1-categories, in which case we should require each type hom A (x, y) to be a *set* in the precise sense of homotopy type theory (i.e., a 0-truncated type, containing no higher homotopy). But what sort of type should A_0 be? The choice we make governs how equality of objects in A get treated.

When A_0 is also a set, then we get a notion of "category" that behaves more like categories in set-theoretic foundations. In such categories, we can compare objects for equality, and this "equality" behaves like equality always has in set theory. This sort of equality of objects is stricter than isomorphism.

Thus, we can have objects that are unequal but isomorphic, and we can perform constructions that violate the principle of equivalence. Another option, however, is to have A_0 be a 1-type (that is, a 1-truncated type, behaving like a 1-groupoid) in which *equality* is identified with *isomorphism* in the category A. More precisely, consider the equality type:

Id A_0 (x, y),, which is defined for any x, y: $A_{0.}$

In homotopy type theory, equality in general types behaves like paths in a space or equivalences in a higher groupoid, so that two points can be equal in more than one-way. In this case, the equality type of A_0 comes with a canonical map id to iso:

Id: A_0 $(x, y) \rightarrow$ iso A (x, y),

into the set of isomorphisms from x to y in the category A.

Define this by "path induction." We say that a category is *saturated* or *univalent* if this map Id to iso is an equivalence, for all objects x and y.

One must emphasize, however, that relative to set theory, it does this by expanding the notion of "equality" to coincide with isomorphism, rather than restricting the notion of isomorphism to coincide with equality. Categorical saturation is where the equality type between objects is identified with the type of *unitary* isomorphisms.

Saturated categories always satisfy the principle of equivalence.

Any construction we can perform in type theory must respect the equality types that are essentially what path induction says. Thus, since in a saturated category, equality of objects is identified with isomorphism, it follows that any construction we can perform must also respect isomorphism of objects. Category set is always saturated. We can furthermore prove that if we take the full image of any category inside its presheaf category, we obtain a "weakly equivalent" category which is saturated, and which is the saturation of the original category.

Saturated categories have other nice properties as well. Axiom of choice does not have to be invoked for all foundational statements. For instance, consider the statement "every fully faithful and essentially surjective functor is an equivalence of categories." In classical set theoretic foundations, this statement is equivalent to the axiom of choice. However, for *saturated* categories, this statement is just *true*, without the need for any axiom of choice. It's essentially a "functor comprehension principle", or a categorical version of the "principle of unique choice": if something is determined up to unique isomorphism, then by univalence it's determined up to unique *equality*, and hence we don't need any axiom of choice in order to choose it.

For our pedgogics might be more illuminating to view Homotopy from models and types, c.f. (Arndt-Kapulkin, 2012) is what was brought to this author's view only before writing this book, during May 2013. The authors attempt to obtain sound models of type theory with the type constructors Π and Σ within the "model-categorical" framework. In good cases that is when some additional coherence conditions (see [GvdB10]) are satisfied, our notion of a model extends the well-known models for the Id- types. Following [Kap10] we propose a set of conditions on a model category that provide enough structure in order to interpret those type constructors. Such a model category will be called a logical model category. Our intention was to give conditions that on one hand will be easy to check but on the other hand, will provide a wide class of examples. It is important to stress that this paper presents only a part of the ongoing project [AK12] devoted to study of Π- and Σ-types in homotopy-theoretic models of type theory.

6.8.1 TYPES AND MODELS

In this section we will review some basic notions of type theory, e.g. Martin-Löf Type Theory is a dependent type theory i.e. apart from simple types and their terms it allows type dependency Hypothesis $\Gamma, x{:}A \vdash B(x)$ type.

In this example B can be regarded as a family of types indexed over A.

There are some basic types as for example: 0, 1, Nat and some type-forming operations. The latter can be divided into two parts:

– simple type-forming operations such as A // B, A × B, and A + B. – operations on dependent types such as Πx:AB(x), Σx:AB(x), and IdA(x, y).

The language of type theory consists of hypotheses of the form

1. $\Gamma \vdash A$ type
2. $\Gamma \vdash A = B$ type
3. Γ□ a A
4. $\Gamma \vdash a=b:A$

There are two more forms, derivable from the ones given above:

5. $\Gamma \vdash \Delta$ cxt
6. $\Gamma \vdash \Delta = \Phi$ cxt

Let us consider for now two basic type consturctors Π and Σ that we can represent. From Curry-Howard isomorphism they correspond to the universal and existential quantification respectively (i.e. ∀ and ∃). Π-types. The version presented below may be different from some other presentations. As the elimination rule we take a weak version that is sometimes called Π-application rule.

$\Gamma, x{:}A \vdash B(x)$ type $\Gamma \vdash \Pi x{:}AB(x)$ type Π-form

$\Gamma, x{:}A \vdash B(x)$type $\Gamma, x{:}A \vdash b(x){:}B(x)$ $\Gamma \vdash \lambda x{:}A.b(x) : \Pi x{:}AB(x)$

$\Gamma \vdash f{:}\Pi x{:}AB(x)$ $\Gamma \vdash a{:}A$ Π-elim $\Gamma \vdash app(f,a){:}B(a)$

Π-intro

$\Gamma, x{:}A \vdash B(x)$type $\Gamma, x{:}A \vdash b(x){:}B(x)$ $\Gamma \vdash a{:}A$ $\Gamma \vdash app(\lambda x{:}A.b(x),a) = b(a)$: B(a)

Π-comp

Σ-types. We use an axiomatization of the Σ-types as an inductive type from the Calculus of Inductive Construction.

$\Gamma \vdash A$ type $\Gamma, x{:}A \vdash B(x)$ type $\Gamma \vdash \Sigma x{:}AB(x)$ type

$\Gamma \vdash A$ type $\Gamma, x{:}A \vdash B(x)$ type $\Gamma, x{:}A, y{:}B(x) \vdash pair(x,y) : \Sigma x{:}AB(x)$ Σ-intro

$\Gamma, z{:}\Sigma x{:}AB(x) \vdash C(z)$ type $\Gamma, x{:}A, y{:}B(x) \vdash d(x,y) : C(pair(x,y))$

Σ-form

$\Gamma, z{:}\Sigma x{:}AB(x) \vdash splitd(z) : C(z)$ $\Gamma, z{:}\Sigma x{:}AB(x) \vdash C(z)$ type

Σ-elim

$\Gamma, x{:}A, y{:}B(x) \vdash d(x,y) : C(pair(x,y))$ $\Gamma, x{:}A, y{:}B(x) \vdash splitd(pair(x, y)) = d(x, y) : C(pair(x, y))$ Σ-comp

6.8.2 WHAT ARE MODEL CATEGORIES

Following is a glimpse from an exposition from a technical report (Arndt-Kapulkin, 2012).

Definition 6.14 A model category is a finitely complete and cocomplete category C equipped with three subcategories: F (fibrations), C (cofibrations), and W (weak equivalences) satisfying the following two conditions:

(Two-of-three) Given a commutative triangle 2. if any two of f, g, h belong to W, then so does the third. 2. Both (C, F ∩ W) and (C ∩ W, F) are weak factorization systems.

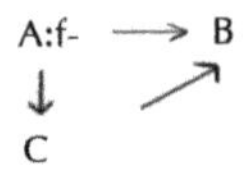

We will refer to model categories sometimes by the tuple (C, W, C, F) or, if no ambiguity arises, just by the underlying category C. We should also mention that some authors add an additional axiom that the classes W, C, and F are closed under retracts.

The following are examples of model categories:

On any complete and cocomplete category C one has the discrete model structure with C:= F:= mor C and W:= iso C. This is the only model structure with W = iso C.

The category Gpd of groupoids has a structure of a model category with: F:= fibrations, C:= functors injective on objects and W:= categorical equivalences

The category sSets:= SetsΔ^{op} of simplicial sets (where Δ is the category of finite nonempty linearly ordered sets) has a standard model structure with W:= {those morphisms inducing isomorphisms on all homotopy groups}, C:= {monomorphisms} and F:= (W ∩ C)⋔

Definition 6.15 An object A is called fibrant if the canonical map A →1 is a fibration. Similarly, an object is called cofibrant if 0 →A is a cofibration.

Proposition 6.2 The following are equivalent for a pair of adjoint functors L: C ⇆ D: R between model categories:

1. L preserves cofibrations and trivial cofibrations. 2. R preserves fibrations and trivial fibrations.

An adjoint pair satisfying these conditions is called a Quillen adjunction and it induces an adjunction Ho(C) $\leftrightarrows$ Ho(D) between the homotopy categories. It is called a Quillen equivalence if this induced adjunction is an equivalence of categories.

6.8.3. GENERIC FUNCTOR MODEL CATEGORIES

Considering the new developments since the author's 1995–2000 publications let us examine how we can present model categories, for example, starting with a Martin Lof types.

The **D<A, G>** category is the category for models definable with D<A, G> and their morphisms. The techniques we are presenting by the three categories save us from having to yet develop a categorical interpretation for $\mathbf{L}_{\omega 1, K}$, in categorical logic as in (Lawvere's, 1967) Chapter 5, and categorical logic, c.f. Chapter 1. A computing view to the functorial models are presented by defining Hasse diagrams on the $L_{\omega l, K}$ fragments. The limit model is defined by computing Hasse diagram limits and natural transformations on the limit cones.

Objects: $L_{\omega l, K}$ Models definable with D<A, G>;

Morphisms: H: <h: M $\rightarrow$ M'>, where h is a model homomorphism.

An embedding from a model definable in *D<A, G>* to the limit created from the functor from the Infinitary language category $L_{\omega l, K}$ to Set can be defined.

Let us carry on with the fragment $\mathbf{L}_{\omega 1, K}$ for the time being. Let ML-denote a Martin Lof Type system with constructors Π and Σ.

Definition 6.16 A generic $L_{\omega 1, K}$ ML-diagram is a diagram ML-definable on $L_{\omega 1, K}$.

Theorem 6.7 There is a generic model functor that creates a ML D<A, G> model.

Proof Follows from the genenric model theorems, the initial model theorem on D<A, G> and ML definability. Cf. exercises.

Proposition 6.3 The generic D<A, G> model is a model category for the ML-types $L_{\omega 1, K}$ definable.

6.9 FILTERED COLIMITS AND COMMA CATEGORIES

The functors from **Set** → **Set** in universal algebra often have the property that they are "determined" by their values on *finite* sets. To be more precise: given any functor **FinSet** → **Set**, there is a canonical way of extending it to a functor **Set** →**Set** (namely, left Kan extension). A functor **Set** → **Set** is called **finitary** if when you restrict down to **FinSet** and then extend back up to **Set** again, you get back the functor that you started with.

For example, the free group functor T(X): **Set** → **Set**, sending a set X to the set T(X) of words in X, is finitary. Informally, this is because the theory of groups involves only *finitary* operations: each operation takes only finitely many arguments. Thus, each element of the free group on X touches only finitely many elements of X. The same is true for any other finitary algebraic theory: rings, lattices, Lie algebras, etc.

So finitary functors are useful. Now the key fact is that a functor from **Set** to **Set** is finitary if and only if it preserves filtered colimits. This immediately suggests that filtered colimits are interesting. This fact is also rather useful: for example, it tells us that the class of finitary functors is closed under composition, which wasn't obvious from the definition. The class of *filtered* colimits can be viewed as the complement of the class of *finite* colimits.

The most general comma category construction involves two functors with the same codomain. Often one of these will have domain **1** (the one-object one-morphism category). Some accounts of category theory consider these special cases only, but the term comma category is actually much more general.

Suppose that **A**, **B**, and **C** are categories, and S and T (for source and target) are functors

A → S→ **C** ←T ¬ **B**

We can form the comma category S ¯ T as follows:

- The objects are all triples (α,β, f) with α an object in **A**, β an object in **B**, and f: S(α)→ T(β) a morphism in **C**.

- The morphisms from (α, β, f) to (α', β', f') are all pairs (g, h) where g: α → α' and h: β → β' are morphisms in **A** and **B** respectively, such that the following diagram commutes:

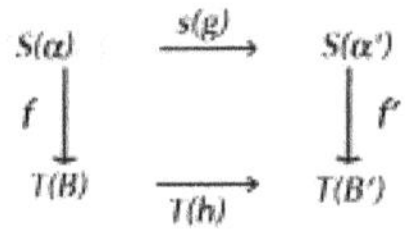

Morphisms are composed by taking (g, h) o (g', h') to be (g o g', h o h'), whenever the latter expression is defined. The identity morphism on an object (α, β, f) is ($id_α$, $id_β$)

Example: Arrow category

S and T are identity functors on C. In this case, the comma category is the arrow C $^{\to}$. Its objects are the morphisms of C, and its morphisms are commuting squares in C.

For each comma category there are forgetful functors from it.

Domain functor, **S ¯ T** → **A**, which maps:

objects: α, β, f → α;

morphisms: (g, h) |→ g

Codomain functor, **S ¯ T** → **B**, S which maps:

objects: α, β, f |→ β;

morphisms: (g, h) |→ h

Important examples with a natural definition in terms of comma categories:

The category of pointed sets is a comma category, (·↓**Set**) with · being (a functor selecting) any singleton set, and **Set** (the identity functor of) the category of sets. Each object of this category is a set, together with a function selecting some element of the set: the "basepoint". Morphisms are functions on sets, which map basepoints to basepoints.

Colimits in comma categories can be "inherited". If **A** and **B** are cocomplete, S: **A** → **C** is a cocontinuous functor, and T: **B** → **C** another functor (not necessarily cocontinuous), then the comma category **S↓T** produced will also be cocomplete. All (small) colimits exist. This result is much harder to obtain directly.

If **A** and **B** are complete, and both S: **A** → **C** and T: **B** → **C** are continuous functors, then the comma category **S** ↓ **T** is also complete, and the projection functors **S** ↓ **T** → **A** and **S** ↓ **T** → **B** are limit preserving.

The notion of a universal morphism to a particular colimit, or from a limit, can be expressed in terms of a comma category. Essentially, we create a category whose objects are cones, and where the limiting cone is a terminal object; then, each universal morphism for the limit is the morphism to the terminal object. This works in the dual case, with a category of cocones having an initial object.

Lawvere showed that the functors F: C ⇒ D and G: D ⇒ C are adjoint if and only if the comma categories F $\downarrow id_D$ and $id_C \downarrow G$ with $\downarrow id_D$ and id_C the identity functors on C and D respectively, are isomorphic, and equivalent elements in the comma category can be projected onto the same element of C *x* D. This allows adjunctions to be described without involving sets, and was the motivation for introducing comma categories.

Another important example is natural transformations:

If the domains of S and T are equal, then the diagram which defines morphisms in S ↓Twith α =α,'β =β,' g=h, is identical to the diagram which defines a natural transformation S⇒ T. The difference being that a natural transformation is a particular collection of morphisms of type of the formS(α)→ T(α), whereas objects of the comma category contains *all* morphisms of type of such form. A functor to the comma category selects that particular collection of morphisms. Succinctly a natural transformation η: S ⇒ T, with S, T: A → C, corresponds to a functor A⇒S ↓T,which mapseach objectαto (α,α, ηα) and maps each morphism h to (h, h).

6.10 MORE ON YONEDA LEMMA

From Section 6.7.1 and Chapter 7-Theorem 7.1, for example, on presheaves and natural transformations that the aurhor hinted on a newer characterizations for Yoneda lemma as indicated on the specific sections, might be helpful to review a paragraph or two on newer disections on the lemma. Modular characterization for the Yoneda lemma might be helpful for a better comprehension on how there can be alternatives to representations considering a new dissection, Pratt (2012) unpublished Staford CS. The Yoneda lemma, algebraic form is as follwos:The internal and external modules of an algebraic structure are isomorphic. A module (M, X, ·) consists of a monoid M = (M, ∘, 1), a set X, and a scalar multiplication $\circ : M \times X \rightarrow X$ satisfying $(m \circ n) \cdot x = m \cdot (n \cdot x)$ and $1 \cdot x = x$. The example is a scalar multiplication of an arbitrary vector space. Notice that any Boolean algebra B determines a module by taking X the underlying set B, M to be the monoid formed by the four-unary Boolean terms under composition. Scalar multiplication the interpretation of the terms as the four-unary operations of B.

Considering representation is applicable to T -algebras or models of an algebraic theory T (*c*.f. chapter 5) defined by equations between terms formed from variables and operation symbols. Typical such structures include monoids, groups, vector spaces, Boolean algebras, lattices, or pointed sets. Every T -algebra A = (A, f1A, f2A,...) is an expansion by nonunary operations of the module (T1, A, ·) where T1 is the monoid of unary terms of T and · is application of these terms interpreted as operations of A. This is internal module of A formed by its unary operations. When all operations of T are unary a T -algebra becomes a module.

The Yoneda Lemma as stated has nothing to say about the homomorphisms between T -algebras, or natural transformations between the presheaves, other than those from objects of J ~= T ^{op}. For a small category, for example, the subcategory of Set consisting of the finite sets is equivalent to a category with objects the natural numbers, however if Set has been constituted to have a proper class of singletons. There is no need to extend the represen- tation of algebras by homomorphisms to the representation of homomorphisms between algebras because they already represent themselves. However, the Yoneda Lemma can be rephrased as a representation theorem for a more abstract notion of presheaf and morphisms, for example, consider Theorem 7.1

6.11 EXERCISES

1. A functor G is a sunfunctor of F on a set S, when every G(S) ⊆ F(S) and every G (f) is a restriction of F(f), for f : S → T, T⊆ S.

Let F be a functor and G0 a function which assigns to each set S a subset G0(S) of F(S). Then G0 is the object function of a unique subfunctor of F iff f: S→ T and s ∈G0(S) implies F(f) s ∈ G0(T).

2. Given a small category **A**, if you freely adjoin finite colimits to **A** and then freely adjoin filtered colimits to that, the end result is the same as if you'd freely adjoined all small colimits to **A**.
3. Let J be a filtered cetgory. Show that there is natural Grothendeik topology on J with filtered colimits.
4. Giraud's axioms for a category *C* are:
   C has a small set of generators, and admits all small colimits. Furthermore, colimits commute with fiber products. Sums in *C* are disjoint. In other words, the fiber product of *X* and *Y* over their sum is the initial object in *C*. All equivalence relations in *C* are effective.
   The last axiom needs the most explanation. If *X* is an object of *C*, an "equivalence relation" *R* on *X* is a map $R \to X \times X$ in *C* such that for any object *Y* in *C*, the induced map Hom(*Y*, *R*) → Hom(*Y*, *X*)×Hom(*Y*, *X*) gives an ordinary equivalence relation on the set Hom(*Y*, *X*). Since *C* has colimits we may form the coequalizer of the two maps *R*→*X*; call this *X*/*R*. The equivalence relation is "effective" if the canonical map $R \to X_{x/R} X$ is an isomorphism.
   (i) Check that
   Giraud's theorem are intantiated on "sheaves on sites" The category of sets is an important special case: it plays the role of a point in topos theory.
   (ii) How is a set a sheaf on a point.
5. Every category with both finite and filtered colimits has all (small) colimits.
6. A ringed topos is a pair *(X, R)*, where *X* is a topos and *R* is a commutative ring object in *X*. (i) Show 2 or more ring space consturctions that are valid for ringed topoi.

(ii) The category of *R*-module objects in *X* is an abelian category with enough injectives.

7. What are stone duality instances on Heyting algebras.
8. Define the natural transformation completing the proof for proposition 6.*x*.
9. The natural transformations on functors sending arbitrary objects to sets on fragment string sets, arrowed by preorder functions, with cones to base F from **{}**, on the above on the fucntor F: $\mathbf{L}_{\omega 1, \mathbf{K}}{}^{\mathrm{op}} \rightarrow$ **Set,** are filter creating representable presheaves for the objects in $\mathrm{L}_{\omega 1, \mathrm{K}}$.
   What is a comma categorical characterization for the above?
10. Prove proposition 6.1.
11. The monics of a subobject in general have many domains, all of which however will be in bijection with each other.
12. The category of graphs is (Set ↓ D), with D: Set ⇒ Set, the functor taking a set S to S *x* S.
    (i) What are the objects?
    (ii) What are the morphisms? When translated, must be the same as the edge for the translated index.
    (iii) Show that in above construction of the category of graphs, the category of sets is cocomplete, nd the identity functor is cocontinuous: so graphs are also cocomplete.

KEYWORDS

- **Deligne-Mumford stacks**
- **functors**
- **power object**
- **powerset**
- **skyscraper sheaf functor**
- **topos**

CHAPTER 7

MODELS, SHEAVES, AND TOPOS

CONTENTS

The category of presheaves on a topological space *X* is a functor category: we turn the topological space into a category *C* having the open sets in *X* as objects and a single morphism from *U* to V if and only if *U* is contained in V. The category of presheaves of sets (abelian groups, rings) on *X* is then the same as the category of contravariant functors from *C* to **Set**. Because of this example, the category Funct(C^{op}, **Set**) is sometimes called the "category of presheaves of sets on *C*" even for general categories *C* not arising from a topological space. To define sheaves on a general category *C*, one needs more structure: a Grothendieck topology on. Categories that are equivalent to **Set** *C* are at times called *presheaf categories*.

7.1 PRESHEAVES

From the preceding chapter above we note that there is contravariant functors.

$F: L^{Op}_{\omega 1, K} \rightarrow$ **Set** to form a presheave.

Lemma 7.1 The limiting cones that are the basis for small completeness for $L_{\omega 1, K}$ define the example representable presheaves for the objects in $L_{\omega 1, K}$.

This natural transformation: example is a category with objects cones to base F from {}, with morphism what corresponds to the fragment preorders. The new functor G on the Op category $\mathbf{L}_w$ is defined based on isomorphisms.

Theorem 7.1 The natural transformations on functors sending arbitrary objects to sets on fragment string sets, arrowed by preorder functions, with cones to base F from **{}**, on the above on the functor F: $L_{\omega 1, K}{}^{op} \rightarrow$ **Set**, are filter creating representable presheaves for the objects in $\mathbf{L}_{\omega 1, \mathbf{K}}$.

A direction from here is what comma category realizes the above.

7.2 DUALITY, FRAGMENT MODELS, AND TOPOLOGY

Starting with basic stone duality we carry on towards functorial models that can be topos characterized with topological structure on fragment models. Points x and y in a topological space X can be separated by neighborhoods if there exists a neighborhood U of x and a neighborhood V of y such that U and V are disjoint ($U \cap V = \emptyset$). X is a Hausdorff space if any two distinct points of X can be separated by neighborhoods. This condition is a third separation axiom, which is why Hausdorff spaces are also called T_2 *spaces*, the name *separated space* is also used.

There is a weaker notion that of a ***preregular space***. X is a preregular space if any two topologically distinguishable points can be separated by neighborhoods. Preregular spaces are also called R_1 *spaces*. A topological space is Hausdorff if and only if it is both preregular: topologically distinguishable points are separated by neighborhoods; and Kolmogorov: distinct points are topologically distinguishable. A topological space is preregular if and only if its Kolmogorov quotient is Hausdorff.

For any Boolean algebra B, $S(B)$ is a compact totally disconnected Hausdorff space; such spaces are called **Stone spaces** (also *profinite spaces*). Conversely, given any topological space X, the collection of subsets of X that are clopen (both closed and open) is a Boolean algebra. Stating that in terms of ultarfilters put us in perspective with Chapter 5. Each Boolean algebra B has an associated topological space, denoted here $S(B)$, called its **Stone space**. The points in $S(B)$ are the ultrafilters on B, or equivalently the homomorphisms from B to the two-element Boolean algebra. The topology on $S(B)$ is generated by a basis consisting of all sets of the form $\{x \in S(B) \mid b \in x\}$, where b is an element of B.

7.2.1 STONE DUALITY

The most general duality referred to as "Stone duality" is the duality between the category ***Sob*** of sober spaces with continuous functions and the category of *Spatial Frames* with appropriate frame homomorphisms. A *sober space* is a topological space with irreducible closed subset of X the closure of exactly one point of X: that is, this closed subset has a unique generic point. Any Hausdorff (T2) space is sober, the only irreducible subsets being points, and all sober spaces are Kolmogorov (T0), and both implications are strict.

Sobriety of X is precisely a condition that forces the lattice of open subsets of X to determine X up to homeomorphism, directing to *pointless topology*. Sobriety makes the specialization preorder a directed complete partial order. Any Hausdorff (T2) space is sober, the only irreducible subsets being points, and all sober spaces are Kolmogorov (T0), with implications that are strict. Sobriety makes the specialization preorder a directed complete partial order. More generally, the underlying topological space of any scheme is a sober space.

The starting point for the mathematics for dualities is the fact that every topological space is characterized by a set of points X and a system $\Omega(X)$ of *open sets* of elements from X, i.e., a subset of the powerset of X. It is known that $\Omega(X)$ has certain special properties: it is a complete lattice within which suprema and finite infima are given by set unions and finite set intersections, respectively. Furthermore, it contains both X and the empty set. Since the embedding of $\Omega(X)$ into the powerset lattice of X preserves finite infima and arbitrary suprema, $\Omega(X)$ inherits the following distributive law: $x \wedge \vee S = \vee \{x \wedge s \colon s \in S\}$, for every element (open set) x and every subset S of $\Omega(X)$. Hence $\Omega(X)$ is not an arbitrary complete lattice but a *complete Heyting algebra*, also called *frame* or *locale*—the names are used to distinguish several categories that have the same class of objects but different morphisms: frame morphisms, locale morphisms and homomorphisms of complete Heyting algebras..

Further example is the category Top of topological spaces that has as morphisms the continuous functions, where a function f is continuous if the inverse image $f^{-1}(O)$ of any open set in the codomain of f is open in the domain of f. Thus any continuous function f from a space X to a space Y defines an inverse mapping f^{-1} from $\Omega(Y)$ to $\Omega(X)$. Furthermore, it is easy to check that f^{-1} (like any inverse image map) preserves finite intersections and arbitrary unions and therefore is a *morphism of frames*. If we define $\Omega(f) = f^{-1}$ then Ω becomes a contravariant functor from the category Top to the category Fram of frames and frame morphisms. Finding a characterization of topological spaces in terms of their open set lattices is equivalent to finding a functor from **Fram** to **Top** which is adjoint to Ω.

7.2.2 POINTS OF A LOCALE

Let us define a functor pt from **Fram** to **Top** that in a certain sense "inverts" the operation of Ω by assigning to each locale *L* a set of points pt(*L*) with a suitable topology. However, we cannot recover the set of points from the locale in general. In general pt does not reproduce all of the original elements of a topological space just from its lattice of open sets. For example all sets with the indiscrete topology yield (up to isomorphism) the same locale, such that the information on the specific set is no longer present. However, there is a reasonable technique for obtaining "points" from a locale that is obtained from a central construction for Stone-type duality theorems.

Consider the points of a topological space *X*. Note that any point *x* gives rise to a continuous function *px* from the one element topological space 1 (all subsets of which are open) to the space *X* by defining *px*(1) = *x*. Conversely, any function from 1 to *X* determines one point: the element that it "points" to. Therefore the set of points of a topological space is equivalently characterized as the set of functions from 1 to *X*.

Now to define a functor pt from **Fram** to **Top** that "inverts" the operation of Ω by assigning to each locale *L* a set of points pt(*L*)— hence the notation pt, with a suitable topology. One cannot expect in general that pt can reproduce all of the original elements of a topological space just from its lattice of open sets—for example all sets with the indiscrete topology yield, up to isomorphism, the same locale, such that the information on the specific set is no longer present. However, there is still a reasonable technique for obtaining "points" from a locale, which indeed gives an example of a central construction for Stone-type duality theorems.

First consider the points of a topological space *X*. One is usually tempted to consider a point of *X* as an element *x* of the set *X*, but any point *x* gives rise to a continuous function *px* from the one element topological space 1 (all subsets of which are open) to the space *X* by defining *px*(1) = *x*. Conversely, any function from 1 to *X* clearly determines one point: the element that it "points" to. Therefore the set of points of a topological space is equivalently characterized as the set of functions from 1 to *X*.

When using the functor Ω to pass from **Top** to **Fram**, all set-theoretic elements of a space are lost, but—using a fundamental idea of category theory one can work direct on the function spaces. Since any "point" px: $1 \to X$ in **Top** is mapped to a morphism $\Omega(px)$: $\Omega(X) \to \Omega(1)$. The open set lattice of the one-element topological space $\Omega(1)$ is just (isomorphic to) the two-element locale $2 = \{ 0, 1 \}$ with $0 < 1$. From these observations becomes reasonable to define the set of points of a locale L to be the set of frame morphisms from L to 2. Yet, we cannot assume that every point of the locale $\Omega(X)$ is in one-to-one correspondence to a point of the topological space X, considering the indiscrete topology, for which the open set lattice has only one "point".

Before defining the required topology on pt(X), let us clarify the concept of a point of a locale further. The perspective motivated above suggests to consider a point of a locale L as a frame morphism p from L to 2. But these morphisms are characterized equivalently by the inverse images of the two elements of 2. Now that a set of points is available for any locale, it remains to equip this set with an appropriate topology in order to define the object part of the functor pt. This is done by defining the open sets of pt(L) as $\varphi(a) = \{ p \in \text{pt}(L) \mid p(a) = 1 \}$, for every element a of L. Here we viewed the points of L as morphisms, but one can state a similar definition for other equivalent characterizations.

Finally pt can be defined on morphisms of **Fram** canonically by defining, for a frame morphism g from L to M, pt(g): pt(M) $\to$ pt(L) as pt(g)(p) $= p \circ g$. In words, we obtain a morphism from L to 2 (a point of L) by applying the morphism g to get from L to M before applying the morphism p that maps from M to 2. Again, this can be formalized using the other descriptions of points of a locale as well:$(p \circ g)^{-1}(0)$.

As noted several times before, pt and Ω usually are not inverses. In general neither is X homeomorphic to pt($\Omega(X)$) nor is L order-isomorphic to Ω(pt(L)). However, when introducing the topology of pt(L) above, a mapping φ from L to Ω(pt(L)) was applied. This mapping is indeed a frame morphism. Conversely, we can define a continuous function ψ from X to pt($\Omega(X)$) by setting h $(x) = \Omega(px)$, where px is just the characteristic function for the point x from 1 to X as described above. At this point we are over reaching that: the functors Ω and pt define an adjunction between the categories **Top** and **Loc** = **Fram**^{op}, where pt is right adjoint to Ω and the natural transformations h and φ^{op} provide the required unit and counit, respectively. For a space X, h: $X \to$ pt($\Omega(X)$) is a homeomorphism if and only if it is bijective.

Finally, one can verify that for every space X, $\Omega(X)$ is spatial and for every locale L, pt(L) is sober. Hence, it follows that the above adjunction of **Top** and **Loc** restricts to an equivalence of the full subcategories **Sob** of sober spaces and **SLoc** of spatial locales. This goal is completed by the observation that for the functor pt o Ω, sending each space to the points of its open set lattice is right adjoint to the inclusion functor from **Sob** to **Top**. For a space X, pt($\Omega(X)$) is called its **soberification**. The case of the functor Ω o pt is symmetric but a special name for this operation is not commonly used. When restricting further to coherent sober spaces ,which are Hausdorff, one obtains the category **Stone** of so-called Stone spaces. On the side of category of distributive lattices, **Dlat**, the restriction yields the subcategory **Bool** of Boolean algebras. Thus one obtains Stone's representation theorem for Boolean algebras.

7.3.3 THE STONE REPRESENTATION THEOREM

One of the most basic constructions of Stone duality: the duality between topological spaces which are *sober* and frames, that is, complete Heyting algebras, which are *spatial. Stone's representation theorem* for *Boolean algebras* states that every Boolean algebra is isomorphic to a field of sets. The theorem is fundamental to the deeper understanding of Boolean algebra that emerged in the first half of the twentieth century. The theorem was first proved by Stone (1936), and thus named in his honor. Stone was led to it by his study of the spectral theory of operators on a Hilbert space.

A simple version of Stone's representation theorem states that any Boolean algebra B is isomorphic to the algebra of clopen subsets of its Stone space $S(B)$. The isomorphism sends an element $b \in B$ to the set of all ultrafilters that contain b. This is a clopen set because of the choice of topology on $S(B)$ and because B is a Boolean algebra.

Restating the theorem with categories; the theorem states that there is a duality between the category of Boolean algebras and the category of Stone spaces. This duality means that in addition to the isomorphisms between Boolean algebras and their Stone spaces, each homomorphism from a Boolean algebra A to a Boolean algebra B corresponds in a natural way to a continuous function from $S(B)$ to $S(A)$. In other words, there is a contravariant functor that gives an equivalence between the categories. The theorem is a special case of Stone duality, a more general framework for dualities between topological spaces and partially ordered sets. The theorem is a special case of Stone duality, a more general framework for dualities between topological spaces and partially ordered sets. The proof requires either the axiom of choice or a weakened form of it. Specifically, the theorem is equivalent to the Boolean prime ideal theorem.

Let us review examples from Boolean algebras to have a closer perspective:

(1) The set $\{0, 1\}$ (with $0 \leq 1$) is a Boolean algebra.

(2) Any complete lattice is a Boolean algebra.

An important notion when reaching to Boolean algebras is first an ideal.

Definition 7.1 Let A be a Boolean algebra. An ideal is a set $I \subset A$ such that

(1) $0 \in I$, (2) if $a \in I$ and $b \in I$, then $a \vee b \in I$, and (3) if $a \in I$ and $b \in A$, then $a \wedge b \in I$.

Say an ideal $I \subset A$ is proper if I/= A, and maximal if it is proper and contains no proper ideals except for itself. Example. Given a topological space X, $\Omega(X)$ (the lattice of open sets ordered by inclusion) is an ideal in $\wp(X)$.

Theorem 7.2 In a Boolean algebra A, an ideal $I \subset A$ is maximal iff for every $a \in A$, either $a \in I$ or $\neg a \in I$, but not both.

Lemma 7.1 (Maximal Ideal Theorem)

Every proper ideal in a Boolean algebra is contained in some maximal ideal.

The basic steps for proving the Stone representation theorem for Boolean algebras is as follows: for example, (Dirk 2011)'s outline.

Definition 7.2 Given a Boolean algebra A, we call S(A) the Stone space associated with A.

Definition 7.3 If X is a Stone space, then the dual algebra of X is the class of clopen sets in X.

Definition 7.4 A field of sets is a Boolean algebra of sets. More formally, take an arbitrary nonempty set X and consider its power set $\wp(X)$. A field of sets is a subset $F \subset \wp(X)$ that is closed under finite set unions, interesections, and complementation.

A field $F \subset \wp(X)$ is separating if, given any distinct $x,y \in X$, there exist disjoint sets $S, T \in F$ such that $x \in S$ and $y \in T$.

Note: There is no connection to the fields of field theory with these notions.

Lemma 7.2 If X is a Stone space and F is a separating field of clopen subsets of X, then F is the dual algebra of X; that is, it is the field of all clopen subsets of X.

Proof We first show that every open set in X can be written as a union of finitely many sets of F. Since F separates points, it also separates points and closed sets.

Theorem 7.3 (Stone Representation Theorem for Boolean Algebras)

Every Boolean algebra is isomorphic to the dual algebra of its associated Stone space.

Let us first view **Stone spaces** from the omitting types persopective, c.f. Chapters 4 and 5. Consider the set of complete *n*-types over *A* as a topological space. Starting with basic stone duality we carry on towards functorial models that can be topos characterized with topological structure on fragment models.

Each Boolean algebra *B* has an associated topological space, denoted here *S*(*B*), called its **Stone space**. The points in *S*(*B*) are the ultrafilters on *B*, or equivalently the homomorphisms from *B* to the two-element Boolean algebra. The topology on *S*(*B*) is generated by a basis consisting of all sets of the form $\{x \in S(B) \mid b \in x\}$, where *b* is an element of *B*.

For any Boolean algebra *B*, *S*(*B*) is a compact totally disconnected Hausdorff space; such spaces are called **Stone spaces** (also *profinite spaces*). Conversely, given any topological space *X*, the collection of subsets of *X* that are clopen (both closed and open) is a Boolean algebra. The most general duality, which is classically referred as "Stone duality" is the duality between the category **Sob** of sober spaces with continuous functions and the category **SFram** of spatial frames with appropriate frame homomorphisms. The dual category of **SFram** is the category of locales denoted by **SLoc**. The categorical equivalence of **Sob** and **SLoc** is the basis for the mathematical area of pointless topology, which is devoted to the study of **Loc**—the category of all locales of which **SLoc** is a full subcategory. The involved constructions are characteristic for this kind of duality, and are detailed below.

Now one can easily obtain a number of other dualities by restricting to certain special classes of sober spaces: The category **CohSob** of coherent sober spaces (and coherent maps) is equivalent to the category **CohLoc** of coherent (or spectral) locales (and coherent maps). On the assumption of the Boolean prime ideal theorem, this statement is equivalent to that assumption. The significance of that stems from the fact that **CohLoc** in turn is dual to the category **DLat** of distributive lattices. Hence, **DLat** is dual to **CohSp**—one obtains Stone's representation theorem for distributive lattices.

When restricting further to coherent sober spaces, which are Hausdorff, one obtains the category **Stone** of so-called Stone spaces. On the side of **DLat**, the restriction yields the subcategory **Bool** of Boolean algebras. Thus one obtains Stone's representation theorem for Boolean algebras. Stone's representation for distributive lattices can be extended via an equivalence of coherent spaces and Priestley spaces (ordered topological spaces, that are compact and totally order-disconnected). One obtains a representation of distributive lattices via ordered topologies: Priestley's representation theorem for distributive lattices.

One of the most basic constructions of Stone duality: the duality between topological spaces which are *sober* and frames (i.e., complete Heyting algebras which are *spatial*. This is a classic achievement of that is based on a deep understanding of the concepts of adjunction, topology, and order theory that are naturally involved.

We revisit Omitting Types from Chapter 4 here towards Stone spaces. Here consider the set of complete n-types over a structure A as a topological space. Consider the following equivalence relation on formulas in the free variables x_1, …, xn with parameters in M:

$$\psi \equiv \phi \Leftrightarrow \mathcal{M} \models \forall x_1, \ldots, x_n(\psi(x_1, \ldots, x_n) \leftrightarrow \phi(x_1, \ldots, x_n)).$$

We can show that $\psi == \varphi$iff they are contained in exactly the same complete types. The set of formulae in free variables $x_1, \ldots, xn$ over A up to this equivalence relation is a Boolean algebra (and is canonically isomorphic to the set of A-definable subsets of Mn). The complete n-types correspond to ultrafilters of this Boolean algebra. The set of complete n-types can be made into a topological space by taking the sets of types containing a given formula as basic open sets. This constructs the Stone space that is compact, Hausdorff, and totally disconnected.

The complete theory of algebraically closed fields of characteristic 0 has quantifier elimination, which allows one to show that the possible complete 1-types correspond to: Roots of a given irreducible nonconstant polynomial over the rationals with leading coefficient 1. For example, the type of square roots of 2. Each of these types is an open point of the Stone space.

Transcendental elements, that are not roots of any nonzero polynomial. This type is a point in the Stone space that is closed but not open.

In other words, the 1-type correspond exactly to the prime ideals of the polynomial ring $\mathbf{Q}[x]$ over the rationals $\mathbf{Q}$: if r is an element of the model of type p, then the ideal corresponding to p is the set of polynomials with r as a root. More generally, the complete n-types correspond to the prime ideals of the polynomial ring $\mathbf{Q}[x1, \ldots, xn]$, in other words to the points of the prime spectrum of this ring.

Given a complete n-type p one can ask if there is a model of the theory that **omits** p, in other words there is no n-tuple in the model which realizes p. If p is an isolated point in the Stone space, that is, if $\{p\}$ is an open set, it is easy to see that every model realizes p (at least if the theory is complete). The **omitting types theorem** says that conversely if p is not isolated then there is a countable model omitting p, provided that the language is countable.

For example, in the theory of algebraically closed fields of characteristic 0, there is a 1-type represented by elements that are transcendental over the prime field. This is a nonisolated point of the Stone space (in fact, the only nonisolated point). The field of algebraic numbers is a model omitting this type, and the algebraic closure of any transcendental extension of the rationals is a model realizing this type. All the other types are "algebraic numbers" (more precisely, they are the sets of first order statements satisfied by some given algebraic number), and all such types are realized in all algebraically closed fields of characteristic 0.

7.3.4 THE SCOTT TOPOLOGY

Scott topology of a space exploits the fact that the operators ▫ and ◊ preserve directed joins. It is now well known in theoretical computer science and topological lattice theory. In more traditional mathematical disciplines, the Scott topology is not as well known, while it appears in real analysis like *semicontinuity in disguise*. One reason that is absent from mathematics courses might be that it is *not Hausdorff*. Whilst there is a compact Hausdorff topology, that one can put on lattices of open sets, it does not have all the properties required. We might have a specific area that such topology can be applied with additional structures to address Topos questions on models for categories.

Definition 7.5 Let $\boldsymbol{L}$ be any complete lattice. A subset $\boldsymbol{U} \subset \boldsymbol{L}$ is called ***Scott-open*** if

it is ***upper***: if $V \sqsupseteq U \in \boldsymbol{U}$ then $V \in \boldsymbol{U}$; and

any subset $\boldsymbol{S} \subset \boldsymbol{L}$ for which $\bigvee \boldsymbol{S} \in \boldsymbol{U}$ already has some *finite* $\boldsymbol{F} \subset \boldsymbol{S}$ with $\bigvee \boldsymbol{F} \in \boldsymbol{U}$.

The Scott-open subsets form a topology on $\boldsymbol{L}$. That is, $\emptyset, \boldsymbol{L} \subset \boldsymbol{L}$ are Scott-open, if $\boldsymbol{U}, \boldsymbol{V} \subset \boldsymbol{L}$ are Scott-open then so is $\boldsymbol{U} \cap \boldsymbol{V} \subset \boldsymbol{L}$, and any union of Scott-open subsets is Scott-open.

The ***compact–open topology*** on the set of continuous functions $X \rightarrow Y$ was introduced by (Fox 1945) Definition 3.1 by Dana Scott identified it as the *crucial* one in the study of topologies on function spaces (Scott 1972) . It had already become clear by then that the *neighbourhoods* of a compact subspace are more important than its *points*. There other examples of the Scott topology that are useful in analysis.

Definition 7.6 A subset $K \subset \mathrm{R}$ obeys the ***constructive least upper bound principle*** if

- it is inhabited and bounded above, and
- for any two real numbers x,z with $x < z$, either z is an upper bound for *all* of K, or there is *some* $k \in K$ with $x < k$.

This condition, which is probably due to L.E.J. Brouwer, is *necessary* to form $y \equiv supK$ because of the *locatedness* property (Axiom 4.9) of y with respect to $(x< z)$, that is, it must satisfy either $x< y$ or $y< z$. That follows from the mixed modal laws and is sufficient to define a Dedekind cut, and therefore a Euclidean real number. The Scott topology is the ***Sierpiński space***, which we call J, is an interesting one. Define it as the lattice of open subspaces of the singleton. Classically, therefore,

having two points and three open sets. Call these points ⊤ and ⊥, the former being open and the latter closed. Since J is a lattice, it also has ∧ and ∨.

The space **2** is both discrete and Hausdorff, but J is neither. Whilst there is a continuous function that takes the two points of **2** to those of J, any continuous function $\vartheta \to \mathbf{2}$ is constant. Hence J is ***connected***, at least in the classical sense.

This means that J has "more than" two points — there is something in between ⊥ and ⊤ that "connects" them. From a constructive point of view, this is because we defined the points of Σ as the open *subsets* of the singleton. There are more of these than just the decidable or complemented ones $\bot \equiv \emptyset$ and $\top \equiv \{\star\}$.

Proposition 7.1. For any space X, there is a bijective correspondence amongst

- open subspaces $U \subset X$,
- continuous functions $\varphi : X \to \vartheta$ and
- closed subspaces $C \subset X$,

where we shall say that φ ***classifies*** $U \equiv \varphi^{-1}(\top)$ and ***co-classifies*** $C \equiv \varphi^{-1}(\bot)$.

In particular, *either U or C uniquely* determines φ.

Notice, therefore, that the correspondence between U and C is given by their common relationship to φ and *not by set-theoretic complementation*. This is how we avoid the double negations that appear frequently in work in the Brouwer and Bishop schools. Nevertheless, it is convenient to retain the word ***complementary*** for this relationship.

In the case where $X \equiv$ J, continuous functions $\vartheta \to \vartheta$ correspond to open subsets of J. Three of these are definable: the identity and the constant functions with values ⊥ and ⊤, corresponding to the singleton, empty and entire open subspaces respectively. Just as there was no arithmetical negation for the ascending reals, *there is no continuous function,* "logical negation", ¬ *that interchanges* ⊥ *and* ⊤.

More generally, Scott-continuous functions respect the order on the lattice. Indeed, *any* topological space X has a ***specialisation order***, defined by $x \sqsubseteq y$ if every neighbourhood of x also contains y. This is antisymmetric iff the space is T_0, discrete iff it is T_1 and (classically) it agrees with the order on the underlying lattice when that is given the Scott topology. Notice that we distinguish this order relation $\sqsubseteq$ from $\leq$ in real and integer arithmetic; they agree in the case of the ascending reals, but $\sqsubseteq$ is $\geq$ or $=$ for the descending or Euclidean reals respectively. The key difference is that the order $\sqsubseteq$ is ***intrinsic***, *i.e.* every continuous function $f{:}X \rightarrow Y$ preserves it, whilst $\leq$ is ***imposed*** on $\mathbb{N}$, $\mathbb{Q}$ and $\mathbb{R}$, in the sense that continuous functions may in general preserve, reverse or ignore it.

Scott continuity is stronger than just preserving order, but instead of talking about arbitrary joins and finite sub-joins, it is convenient to introduce a new definition.

Definition 7.7 A poset (partially ordered set) $(\boldsymbol{I},\leq)$ is ***directed*** if it is inhabited (has an element) and, for any $i,j \in \boldsymbol{I}$, there is some $k \in \boldsymbol{I}$ with $i \leq k \geq j$. When we form a join or union indexed by $\boldsymbol{I}$ (taking $\leq$ to $\sqsubseteq$), we use an arrow to indicate that it is directed: $\bigvee^{\uparrow}$ or $\bigcup^{\uparrow}$.

Example: The set of finite subsets of any set, with the inclusion order.

Proposition 7.2. A function $F{:}\boldsymbol{L}_1 \rightarrow \boldsymbol{L}_2$ between complete lattices is Scott-continuous, *i.e.* $F^{-1}(V)$ is Scott-open in $\boldsymbol{L}_1$ whenever $V \subset \boldsymbol{L}_2$ is Scott-open in $\boldsymbol{L}_2$, iff F preserves directed joins.

Suppose, for example, that we want to find an upper bound for a function $f{:}K \rightarrow \mathrm{R}$. The subsets $Uu \equiv \{k \in K \mid fk < u\}$ indexed by candidate bounds $u \in \mathrm{Q}$ are open and cover K, so only finitely many of them are needed. Now, u ranges over a (totally ordered and so) directed poset Q, and we have $u \leq v \Rightarrow Uu \subset Uv$. Therefore the *finite* open sub-cover need only have *one* member, named by the greatest u in the finite set, and we have $K \subset Uu$ for a single u. In other words, there is a ***uniform*** bound.

Notation Since open sets $U \subset X$ correspond to continuous maps $X \to J$ we write ϑX for the lattice of them, equipped with the Scott topology. This correspondence also gives rise to the notation $\varphi\ a$ or $\varphi\ a \Leftrightarrow \top$ for $a \in U$, for membership of this subspace.

Proposition 7.3 The function *ev*: $\vartheta X \times X \to \vartheta$ is jointly continuous (with respect to the Tychonov product topology defined from the given topology on X and the Scott topology on ϑX and ϑ) iff X is *locally compact* .

As we are dealing with non-Hausdorff spaces (in particular ϑX) here, we need to alter the traditional definition of local compactness.

Definition 7.8 A, not necessarily Hausdorff) space, X is ***locally compact*** if, whenever $x \in U \subset X$ with U open, there are compact K and open V with $x \in V \subset K \subset U$.

This relation between open subsets, written $V \ll U$ and called ***way below***, may be characterised without mentioning the compact subspace K between them: if $U \subset \bigcup^{\uparrow} Wi$ then already $V \subset Wi$ for some i.

The result that justifies calling ϑX a ***function-space*** is then

Theorem 7.4 Let X be locally compact and Γ any space. Then ϑX is also locally compact and there is a bijection between continuous functions

Overtness plays a key role underlying all of this. In the previous section, we saw manifestations of it in open maps, the existential quantifier and the join-preserving property of $\lozenge$. It is also related to recursive enumerability, and specifies *which* joins exist in open set lattices and the ascending real's. However, despite the fact that it does so many jobs, we don't have to do anything to encode this behavior into the system: it will all just fall out naturally.

7.3 LIFTS ON TOPOS MODELS ON CARDINALITIES

Here we introduce new functors on topos to address specific areas on model theory where we can have a sense on how the topological structures can interact. The stance is to present functors from fragment topos with product topologies to topological structures that are obtained from, for example, from a Stone representation on a Scott topology. Example application area is cardinality on models at the Löwenheim–Skolem theorem, named for Leopold Löwenheim and Thoralf Skolem, states that if a countable first-order theory has an infinite model, then for every infinite cardinal number κ it has a model of size κ. The result implies that first-order theories are unable to control the cardinality of their infinite models, and that no first-order theory with an infinite model can have a unique model up to isomorphism.

The (downward) Löwenheim–Skolem theorem is one of the two key properties, along with the compactness theorem, that are used in Lindström's theorem to characterize first-order logic. In general, the Löwenheim–Skolem theorem does not hold in stronger logics such as second-order logic.

The modern statement of the theorem is both more general and stronger than the version for countable signatures stated in the introduction.

In its general form, the *Löwenheim–Skolem Theorem* states that for every signature σ, every infinite σ—structure M and every infinite cardinal number $\kappa \geq |\sigma|$, there is a σ-structure N such that $|N| = \kappa$ and

if $\kappa < |M|$ then N is an elementary substructure of M;

if $\kappa > |M|$ then N is an elementary extension of M.

The theorem is often divided into two parts corresponding to the two points above. The part of the theorem asserting that a structure has elementary substructures of all smaller infinite cardinalities is known as the downward Löwenheim–Skolem Theorem. The part of the theorem asserting that a structure has elementary extensions of all larger cardinalities is known as the upward Löwenheim–Skolem Theorem.

The statement given in the introduction follows immediately by taking M to be an infinite model of the theory. The proof of the upward part of the theorem also shows that a theory with arbitrarily large finite models must have an infinite model; sometimes this is considered to be part of the theorem.

For historical variants of the theorem, see the notes below. Examples and consequences are as follows.

Let **N** denote the natural numbers and **R** the real's. It follows from the theorem that the theory of (**N**, +, ×, 0, 1) (the theory of true first-order arithmetic) has uncountable models, and that the theory of (**R**, +, ×, 0, 1) (the theory of real closed fields) has a countable model. There are, of course, axiomatizations characterizing (**N**, +, ×, 0, 1) and (**R**, +, ×, 0, 1) up to isomorphism. The Löwenheim–Skolem theorem shows that these axiomatizations cannot be first-order. For example, the completeness of a linear order, which is used to characterize the real numbers as a complete ordered field, is a nonfirst-order property.

A theory is called **categorical** if it has only one model, up to isomorphism. This term was introduced by Veblen (1904). The Löwenheim–Skolem theorem dealt a first blow to this hope, as it implies that a first-order theory, which has an infinite model, cannot be categorical. Later, in 1931, the hope was shattered completely by Gödel's incompleteness theorem.

Many consequences of the Löwenheim–Skolem theorem were counterintuitive to logicians in the early twentieth century, since the distinction between first-order and nonfirst-order properties was not yet understood. One such consequence is the existence of uncountable models of true arithmetic, which satisfy every first-order induction axiom but have non-inductive subsets. Another consequence that was considered particularly troubling is the existence of a countable model of set theory, which nevertheless must satisfy the sentence saying the real numbers are uncountable. This counterintuitive situation is known as Skolem's paradox; it shows that the notion of countability is not absolute.

The first significant result in what later became model theory was *Löwenheim's theorem* in Leopold Löwenheim's publication "Über Möglichkeiten im Relativkalkül» (1915): For every countable signature S, every σ-sentence which is satisfiable is satisfiable in a countable model.

Löwenheim's paper was actually concerned with the more general Peirce–Schröder calculus of relatives (relation algebra with quantifiers). He also used the now antiquated notations of Ernst Schröder. For a summary of the paper in English and using modern notations see Brady (2000, Chapter 8).

Skolem (1920) gave a (correct) proof using formulas in what would later be called *Skolem normal form* and relying on the axiom of choice: Every countable theory, which is satisfiable in a model *M*, is satisfiable in a countable substructure of *M*.

Skolem (1923) also proved the following weaker version without the axiom of choice:

Every countable theory, which is satisfiable in a model, is also satisfiable in a countable model. Skolem (1929) simplified Skolem (1920). Finally, Anatoly Ivanovich Maltsev (1936) proved the Löwenheim–Skolem theorem in its full generality (Maltsev, 1936). He cited a note by Skolem, according to which the theorem had been proved by Alfred Tarski in a seminar in 1928. Therefore the general theorem is sometimes known as the *Löwenheim–Skolem–Tarski theorem*. But Tarski did not remember his proof, and it remains a mystery how he could do it without the compactness theorem.

7.3.1 TOPOS GLIMPSE ON LIJWENHEIM-SKOLEM

A difficult theorem of Morley says that if a theory T is countable and κ-categorical for some uncountable cardinal κ, then it is κ-categorical for all uncountable cardinals κ. This shows that the class of theories to which Vaught's test applies is rather limited. Many important complete theories are not κ-categorical for any κ. This limitation of Vaught's test will be partly overcome by means of saturated models.

The standard proof deploys the axiom of choice. The strategy for a topos characterization that applies, for example, product topology on the Kiesler fragment on the preceding sections on topoi, where applying Scott-open topology to obtain Stone represenations on the fragment topoi glimpse direct topoi on reaching Löwenheim–Skolem.

Example applications for the upward Lo¨wenheim-Skolem-Tarski theorem is stated as follows: Let A be an infinite structure. Let κ be a cardinal ≥ max(‖A‖, |sig(A)|). Then A has a proper elementary extension of power κ.

Outline:

Conside the fragment topology defined on the author defined on Kiesler $L_{\omega 1,K}$ fragments. Let us define a discrete topology on the Keisler fragment K, on $L_{1.}$. Let M be the infinite product copies on K. Give K the product toplogy. Let F = {K, K^2,...}.

Observations: (i)A subset of elements of F from a topological space with a pointset topology. (ii) M is homomorphic to its product with itself.

Definition 7.19 If A M $^{n+1}$ the fragment projection of A is { <1, ..., n, > A)}

Theorem 7.5 On Fragment consistency theorem, consider the interim fragment models Mi. M |= iff Mi |= a projection on fragments sets that positively locally realize is the set modeled at the ith tower iteration.

The new areas are as follows:

Theorem 7.6 There is an ultrafilter $I_{F\text{ such}}$ that the fragment model ultra-prodcut UAi is a model of $L_{P,\omega}$. iff Ai |= a projection on fragments sets that Horn locally realize Ai.

Proof The latter condition guarantees that the Ai's that the ultrafilter condition is met on U at Los theorem:

Definition 7.20 Let us say that model M is positively k -saturated if for every X ⊆ |M| of cardinality < k, every type is positively locally realized in M

Corollary 7.1 The interim models Mi, theorem, are positively k -saturated at the interim language fragment.

Corollary 7.2 Each fragment model has,

(i) a k+ saturated elementary extension that is the Ultraproduct on fragments k-saturated at the positive interim language.
(ii) The language has a model iff the iff every finite fragment group elementary extensions have a reduced product filter to the fragment model projections.

Define a functor F from the category on the toplogical structures on the K-fragment to a Scott topology on Boolean models isomorphic to the fragment models.

Lemma 7.1 There is a topological space on $L_{\omega 1, K}$ fragments, with set of points *X* and a system Ω(*X*) of open sets of elements from *X*, that is, a subset of the powerset of *X*.

Proof For what we had shown, there is Grorthendiek topology on the fragment, rest follows form the positive forcing basics on the open sets of elements from the Lw, K fragments, that are subset of the powerset T*.

Lemma 7.2 Starting with the discrete topology on the Keisler fragment K, on L_{1}. Let M be the infinite product copies on K. Give K the product topology. Let F = {K, K^2, …}.

Observations: (i)A subset of elements of F from a topological space with a pointset topology. (ii) M is homomorphic to its product with itself . There is a functor from the K-fragment category to the category of Boolean models with a Stone representation.

Recall that: Let I be a non-empty set. A proper filter U over I is a set of subsets of I such that:

(i) U is closed under supersets; if X ∈ U and X ⊆ Y ⊆ I then Y ∈ U. (ii) U is closed under finite intersections; if X ∈ U and Y ∈ U then X ∩Y ∈ U. (iii)I∈U but ∅∈/U.

An ultrafilter over I is a proper filter U over I such that: (iv) For each X ⊆ I, exactly one of the sets X, I\X belongs to U. We take the the fragment models and projection pointsets on the product topology above. The following proposition shows that the points ultrafliters.

Proposition 7.4 From the solution set theorem adjuction (author ECCT 1996) on model diagram categories there is a forgetful functor form the fragment diagram models to the points on the product topology above. The points are representation ultrafilters.

Proposition 7.5 There is a natural transformation that maps stone representation category for the Boolean models above to the $L_{\omega 1, K}$ fragment models.

We can apply the sheaf topos on authors ASL with forgetful functors on fragments. Forgetful functors to **Set** are often representable. In particular, a forgetful functor is represented by (A, u) whenever A is a free object over a singleton set with generator u.

The forgetful functor **Top** → **Set** on the category of topological spaces is represented by any singleton topological space with its unique element.

To prove proposition 7.5 define two functors:

F1: $L_{\omega 1, K} \rightarrow$ Set

F2: a forget functor: Top → ΠK

where the singleton element is the discrete product topology on the Keisler fragment K.

F2 : TOP ⟶ ΠK
h1 h2
F1 : $L_{\omega 1,k}$ ⟶ Set

h1 is the natural transformation functor from K products to $L_{\omega 1, K}$; h2 is projections to fragment sets.

Theorem 7.7 (Topos Upward Lowenhiem Skolem) k-cardinal realization models are at the k-saturation point set models from the above proposition.

Theorem 7.8 Topos downward Lowenhiem Skolem realization follows from the projective set models on fragments (Author ASL-Sofia and ASL 2013, Portugal).

Proofs combine the elementary diagram method (used in the proof of the previous theorem) with the proof of the Lowenheim-Skolem-Tarski theorem with the following observation:

Consider a scott topology on fragment consistent algebraic models. The fragment models have a Scott-open^{op} top. Then we can apply stone duality to the above topos to lower models.

7.3.2 FILTERS ON CATEGORIES

In category theory, **filtered categories** generalize the notion of directed set understood as a category, hence called a directed category; while some use directed category as a synonym for a filtered category.

A category J is **filtered** when it is not empty,

- for every two objects J and j in J there exists an object k and two arrows f: j → k and f': j' → k in J, for every two parallel arrows u, j: i →j in J there exists an object k and an arrow w: j → k such that wu= wv.

An arrow diagram is said to be of cardinality k if the morphism set of its domain is of cardinality k. A category J is filtered if and only if there is a cone over any finite diagram d: D → J more generally, for a regular cardinal k, a category J is said to be k-filtered if for every diagram d in J of cardinality smaller than k there is a cone over d.

A **filtered colimit** is a colimit of a functor F: J → C where J is a filtered category. This readily generalizes to k-filtered limits. An **ind-object** in a category is a presheaf of sets C^{op} → Set, which is a small filtered colimit of representable presheaves. Ind-objects in a category C form a full subcategory Ind © in the category of functors C^{op} → Set.

Cofiltered Categories: There is a dual notion of **cofiltered** category. A category is cofiltered if the opposite category J^{op} is filtered. In detail, a category is cofiltered when it is not empty and

- for every two objects j and j' in J there exists an object k and two arrows
- f: k → j and f': k → j' in J, for
- for every two parallel arrows arrows u, v: j →I in J

there exists an object k and an arrow w: k → j such that uw=vu.

A **cofiltered limit** is a limit of a functor F: J → C where J is a cofiltered category.

7.3.3 FILTERED COLIMITS

The functors from **Set** → **Set** in universal algebra often have the property that they are "determined" by their values on *finite* sets. To be more precise:

given any functor **FinSet⇒Set**, there is a canonical way of extending it to a functor **Set⇒Set** (namely, left Kan extension). A functor **Set** → **Set** is called **finitary** if when you restrict down to **FinSet** and then extend back up to **Set** again, you get back the functor that you started with.

For example, the free group functor T(X): **Set** ⟶ **Set**, sending a set X to the set T(X) of words in X, is finitary. Informally, this is because the theory of groups involves only *finitary* operations: each operation takes only finitely many arguments. Thus, each element of the free group on X touches only finitely many elements of X. The same is true for any other finitary algebraic theory: rings, lattices, Lie algebras, etc.

So finitary functors are useful. Now the key fact is that a functor from **Set** to **Set** is finitary if and only if it preserves filtered colimits. This immediately suggests that filtered colimits are interesting. This fact is also rather useful: for example, it tells us that the class of finitary functors is closed under composition, which wasn't obvious from the definition. Has been obsereved that class of *filtered* colimits can be viewed as the complement of the class of *finite* colimits.

On the newer developments with the present author's Peter Johnston (2010) has been prominent on newer and the (1977,2002) long time the standard compendium on topos theory. However, he purported to state that his book on Topos that the author has not had the opportunity to examine yet, is very hard to read.

From his students and newer alumni O. Caramello (2012), presents a topos-theoretic approach to Stone-type dualities; several known dualities are seen to be instances of just one topos-theoretic phenomenon, and new dualities are introduced. In fact, infinitely many new dualities between preordered structures and locales or topological spaces can be generated through our topos-theoretic machinery in a uniform way. We then apply our topos-theoretic interpretation to obtain results connecting properties of preorders and properties of the corresponding locales or topological spaces, and we establish adjunctions between various kinds of categories as natural applications of our general methodology.

Several known dualities are seen to be instances of just one topos-theoretic phenomenon, and new dualities are introduced. In fact, infinitely many new dualities between preordered structures and locales or topological spaces can be generated through our topos-theoretic machinery in a uniform way.

Topos-theoretic interpretations are applied to obtain results connecting properties of preorders and properties of the corresponding locales or topological spaces, establishing adjunctions between various kinds of categories as natural applications. In the last part of the paper, we exploit the theory developed in the previous parts to obtain a topos-theoretic interpretation of the problem of finding explicit descriptions of models of ordered algebraic theories. Such latter characterizations were also presented by the present author's publications since 1996, beginning with ECCT-Tours.

7.4 EXERCISES

1. A functor G is a subfunctor of F on a set S, when every G(S) ⊆ F(S) and every G (f) is a restriction of F(f), for f: S ⟶ T, T⊆ S. Let F be a functor and G0 a function which assigns to each set S a subset G0(S) of F(S). Then G0 is the object function of a unique subfunctor of F iff f: S⟶ T and s ∈ G0(S) implies F(f) s ∈ G0(T).
2. Given a small category **A**, if you freely adjoin finite colimits to **A** and then freely adjoin filtered colimits to that, the end result is the same as if you'd freely adjoined all small colimits to **A**.
3. Let J be a **filtered** category. Show that there is natural Grothendeik topology on J with filtered colimits.
4. Show the correspondence between the points in *S*(*B*) ultrafilters on *B*, and the homomorphisms from *B* to the two-element Boolean algebra.
5. Every category with both finite and filtered colimits has all (small) colimits.
6. A ringed topos is a pair *(X, R)*, where *X* is a topos and *R* is a commutative ring object in *X*.
   (i) Show 2 or more ring space constructions that are valid for ringed topoi.
   (ii) The category of *R*-module objects in *X* is an abelian category with enough injectives.
7. What are stone duality instances on Heyting algebras.
8. Define the natural transformation completing the proof for proposition 6.x.

9. Show that an infinite set with the cofinite topology is a T1 space, which is not sober.
10. Consider Theorem 7.2: The natural transformations on functors sending arbitrary objects to sets on fragment string sets, arrowed by preorder functions, with cones to base F from {}, on the above on the functor F: $L_{\omega 1, K}{}^{op} \rightarrow$ Set, are filter creating representable presheaves for the objects in $L_{\omega 1, K}$. What comma category might realize that above.
11. Show that an infinite set with the cofinite topology is not Sober.
12. Topological spaces, which are *sober* and frames are complete Heyting algebras which are *spatial.*
13. From section 7.3.2 Show that setting $\Omega(\mathrm{pt}(L)) = \{\varphi(a) \mid a \in L\}$ yields a topological space $(\mathrm{pt}(L), \Omega(\mathrm{pt}(L)))$. It is common to abbreviate this space as $\mathrm{pt}(L)$.
14. Prove that A function $F{:}\boldsymbol{L}_1 \rightarrow \boldsymbol{L}_2$ between complete lattices (or dcpos) is ***Scott-continuous***, *i.e.* $F^{-1}(V)$ is Scott-open in $\boldsymbol{L}_1$ whenever $V \subset \boldsymbol{L}_2$ is Scott-open in $\boldsymbol{L}_2$, iff F preserves directed joins, *i.e.*
    $F(\bigvee i_{\in} I^{\uparrow}\, xi) = \bigvee i_{\in} I^{\uparrow}\, F(xi)$
    for all directed (xi)i∈ I⊂L1. Hence a function that is Scott-continuous in each of several variables is jointly continuous in them.
15. Prove that Scott topology may be defined on any directed-complete partial order, *i.e.* a poset in which every *directed* subset but not necessarily every finite subset has a join.
16. Let C be an equational category and D ⊆ C be a full subcategory. A necessary and sufficient condition that D be the category of models of a generalized equational Horn theory based on the operations of C is that D is closed under subobjects and products. If the theory of C is finitary, then D is the category of models for a Horn theory if and only if it is also closed under filtered colimits.

KEYWORDS

- **compact–open topology**
- **Grothendieck topology**
- **Hausdorff space**
- **isomorphisms**
- **morphism of frames**
- **preregular space**
- **presheaf**
- **Stone spaces**

CHAPTER 8

FUNCTORS ON FIELDS

CONTENTS

8.1 INTRODUCTION

Fields are an important basis to algebraic coding and cryptography. Specifically, Galois fields have played an important part as a mathematical basis. This chapter addresses fields and prime factorization on polynomials with new model-theoretic techniques. Computationally appealing models, called Prime models are applied to create certificates for models and allow forming product models based on model certificates to provide a basis to encoding and factorization. Algebraically closed fields are presented and applied to polynomial fields that with irreducible factorization can form a basis to cryptography with model certificates. Section 8.2 introduces the new mathematical logic models basics. Section 8.3 presents fields, polynomial factorization, and algebraically closed fields. Section 8.4 presents Prime models and their properties that allow a characterization with model certificates. New theorems are stated on model certifications. Section 8.5 introduces further foundational techniques on fragment consistent models that are applied to prove new model-theoretic consistency and completeness bases. Important foundations on model completions, and filter are developed to allow us to do a model factorization comparable to number-theoretic prime factorization on fields. Section 8.6 develops the new model factorization with specific theorems based on Section 8.5. Models and new techniques on fragment consistent algebras are presented to present models with creative certificates. New areas on models basis to cryptography on algebraic fields were presented by our mathematical sciences venture group elsewhere.

8.2 BASIC MODELS

A proof to Godel's completeness theorem Henkin style proceeds by constructing a model directly from the syntax of the given theory. The structure is obtained by putting terms that are provably equal into equivalence classes, then defining a free structure on the equivalence classes. The computing enterprise requires more general techniques of model construction and extension, since it has to accommodate dynamically changing world descriptions and theories. The techniques the author has applied since 1980s for model building as applied to computing allows us to build and extend models with generic diagrams defined from a minimal set of function symbols with which a model can be defined inductively. Generic diagrams allow us to define canonical models with specific functions. Sentential logic is the standard formal language applied when defining basic models. The language L is a set of sentence symbol closed by finite application of negation and conjunction to sentence symbols. Once quantifier logical symbols are added to the language, the language of first order logic can be defined. A Model A for a language is a structure with a set A, for each constant symbol in the language there corresponds a constant in A. The signature or similarity type for a set of formulas or terms is the set of nonlogical symbols appearing. A languages signature encompasses the arity-coarity of the function and relation symbols.

For each function symbol in the language there is a function defined on A; and for each relation symbol in the language there is a relation defined on A. Godel's incompleteness theorem had in particular spurred explorations on the model theories and fields numbers on Tarski, Kochen, Macintire and Ershov's publications, for example. For the specific areas that interest us, however, the important areas are algebraically closed fields and the field models. From chapter 4 we have the following.

Definition 8.1 Let (M, a)*c* in C be defined such that M is a structure for a language L and each constant *c* in C has the interpretation a in M. The mapping $c \rightarrow$ ac is an assignment of C in M. We say that (M, a)*c* in C is canonical model for a presentation P on language L, iff the assignment $c \rightarrow$ a maps C onto M, that is, M=(a:*c* in C).

Generic diagrams allow us to characterize certain canonical models unique up to isomorphism and to address computability on models based on specific functions. Certain models, called prime models, are applied in the forthcoming sections to address models to cryptographic computing and algebraic coding theory.

Definition 8.2 Let *A; B* be structures with same signature and $|A| \subseteq |B|$. Write $A <$ e B if for all formulas *'(x1; : : : ; xn)* with only the free variables shown, and all *a1; : : : ; an* $\in |A|$, $A \models$*(a1; : : : ; an)* if and only if $B \models$ (a1; *: : : ; an).* We then say that *A* is an *elementary substructure* of *B,* or *B* is an *elementary extension* of *A*. We say that *A* is a *substructure* of *B,or B* is an *extension* of *A, if the above condition holds for all atomic formulas.*

Definition 8.3 Let A; B be structures with same signature. We say that f : |Aj| → |B| is an elementary embedding or elementary monomorphism if f is one-one and,for all formulae '(x1; : : : ; xn) and all a1; : : : ; an ∈ |A| , A ⊨ f(a1; : : : ; an) if and only if B ⊨ (f(a1); : : : ; f(an)).

If the above equivalence holds for atomic 'formulas f called is an embedding or monomorphism of A into B, or an isomorphism of A into (not necessarily onto) B. The concepts of monomorphism and elementary monomorphism are essentially are basic forms of extension and elementary extension.

A set S of sentences is said to be consistent if it has a model. The following theorem is due to Go¨del 1929 in the countable case and Malcev 1936 in the uncountable case.

THEOREM 8.1 (COMPACTNESS THEOREM)

If every finite subset of S is consistent, then S is consistent.

Corollary 8.1 A theory T has a model iff every finitely axiomatize part of T has a model.

8.3 FIELDS

8.3.1 GALOIS FIELDS

Algebraic cryptography and l coding theory have a considerable amount based on finite fields. Galois fields are fields that have finite elements. GF(q) is a specific field with n elements, where n is called the specific Galois field order. For example, for any prime p, the nonnegative integers less than p form a field using module p arithmetic. An expression of the form $f_n D_n + f_{n-1} D_{n-1} + ... + f_0$, denoted f(D), is called a polynomial over GF(q) with degree n, if the coefficient fn is nonzero.

Amongst the important bases is the unique factorization theorem: we state the theorems without proof when are not due to this author and are from the well-established preceding mathematics.

Theorem 8.2 A polynomial f(D) over a given field has a unique factorization into field elements times a product of monic irreducible polynomials over the field, each of degree at least 1.

As an application, suppose f(D) is an irreducible polynomial of degree n over a finite field GF(q). Consider the set of all polynomial of degree n-1 or less over GF(q). Let operation *x* among these polynomials be the remainder of polynomial product modal f(D). Under polynomial addition and multiplication, the set of polynomials g(t) over GF(q) of degree n-1 or less form a field.

Theorem 8.3 Each Galois field has a unique subfield with a prime order.

8.3.2 ALGEBRAICALLY CLOSED FIELDS AND POLYNOMIALS

Definition 8.4 A field *F* is *algebraically closed* if every nonconstant *f(x)* $\in F[x]$ has a root in *F,* i.e. $f(a) = 0$ for some $a \in |F|$. For any field *A,* an *algebraic closure* of *A* is a field extension $B \supseteq A$ such that (i) *B* is algebraic over *A,* and (ii) *B* is algebraically closed.

Proposition 8.1 Every field *A* has an algebraic closure.

To have a concrete glimpse on algebraically closed fields let us examine basic field extensions. Let *A* be a field. A field *extension* of *A* is an extension $B \supseteq A$ such that *B* is a field. A field extension of *A* is basically the same as a model of(filed axioms)∪(diagram of A).If A and B are fields,A⊆B, and b Î |B|,let A[b] be the smallest substructure of *B* containing $|A| \cup \{b\}$, i.e. the subring generated by $|A| \cup \{b\}$. Let *A(b)* be.

A field extension of *A* is said to be *simple* if it is the smallest subfiled of *B* containing $|A| \cup \{b\}$, and is denoted by *A(b).* Let *A* be a field and consider terms *t(x)* with sigature defined by the generic diagram of A and *x* is the only free variable in $t(x)$. Two such terms $t1(x)$ and $t2(x)$ are said to be *equivalent* iff the field axioms) U (diagram of A) $\models \forall x\,(t1(x) = t2(x))$ The set of equivalence classes is a commutative ring and is denoted by $A <x>$.

A polynomial is a term f(x) as above with either f(x)≡ 0 or $f(x) = a_n x_n + \ldots + a_1 x + a_0$, $a_i \in |A|$. Clearly each equivalence class in $A <x>$ contains a polynomial. Each equivalence class contains only one polynomial. Thus the polynomials are a system of representatives for the equivalence classes in $A <x>$. From domain theory basics the polynomials over A form a domain, and any domain is embeddable in a fieldof *quotients.* Thus the set of polynomials over A is embedded in a field extension of A. Hence distinct polynomials represent distinct equivalence classes in $A <x>$.

8.4 PRIME MODELS

Prime models, for example, (Knight 1978; Vaught 1958, 1962) have useful computational properties that had interested the author before the specific considerations here. A new mathematical basis is developed to have specific models for certain algebraically closed fields. Amongst the known appealing Prime model properties are what Vaught theorems had put forth. Taking on countable models there are the following:

Definition 8.5 Let T be a complete theory. A model A of T is said to be *prime* if every model of T has an elementary submodel, which is isomorphic to A.

Theorem 8.4 Let T be a countable complete theory. If T has a prime model, it is unique up to isomorphism.

For example, T = A(p), the theory of algebraically closed fields of prime characteristic p. The prime model is the algebraic closure of (Fp; +, –; 0,1). 2. Let A be any field and let T be an algebraically closed field ∪ (diagram of A), where ∪ is set union. T is complete. The prime model of T is A, the algebraic closure of A.

Definition 8.6 Let Σ be a set of formulas in the variables x1…xn. Let R be a model for L. We say that R realizes Σ iff some n-tuple of elements of A satisfies Σ in R. R omits Σ iff R does not realize Σ.

Let $\sum_n$, n a nonnegative integer, be a set of formulas $\varphi(x_1 \ldots x_n)$ on the language signature $\sum$ for a theory T with no free variables except for $x_1 \ldots x_n$.

Definition 8.9 An n-type over T is a set $\sum_n$, which is consistent over T and closed under logical consequence over T.

Definition 8.10 An n-type f is called principal over T if it is generated by a single formula. That is there exists $\varphi \in \sum_n$ such that $f = \{ \psi \in \sum_n : T \models \forall x_1 \ldots x_n (\varphi(x_1 \ldots x_n) \rightarrow \psi (x_1 \ldots x_n))\}$. φ is called the generator for f.

Theorem 8.5 Let *T* be a countable complete theory. The following are equivalent.

(i) *T* has a prime model.
(ii) *T* has an atomic model.
(iii) Every principal *n-type* over *T* is included in a complete principal *n-type* over *T*.

Theorem 8.6 (Vaught)

The Following are equivalent:
(i) R is a countable atomic model.
(ii) R is a prime model.
(iii) R is a countably prime model.

Theorem 8.7 (Vaught's test for completeness)

Let T be a countable theory such that
(i) T has no finite models;
(ii) For some infinite cardinal κ, T is κ-categorical.
Then T is complete.

8.5 OMITTING TYPES ON FIELDS

Recall from chapter 4 on models: Let p be an n-type over a theory T. A model A of T is said to omit p if there is no n-tuple a1,...,an $\in |A|$ which realizes p.

Notation to fix The class of all models of a theory we denote that by Models(S). A sentence τ is said to be a logical consequence of S ,written $S \models \tau$ if $\Sigma_\tau \subseteq \Sigma_S$ and $M \models \tau$ for all $\in$ M Models (S). We have already seen how to use the compactness theorem to construct a model, which realizes p. It is more difficult to construct a model, which omits p. Indeed, such a model may not even exist. The omitting types theorem gives a sufficient condition for the existence of a model of T, which omits p. If T is complete, this sufficient condition is also necessary. This area is further treated in chapter 9. Here we will not spend more time far afield.

Example application of the omitting types to set theory is the following.

Theorem 8.8 Let A = (|A|, $\in$A) be a countable model of ZF set theory. Then A has a proper elementary extension B such that $\omega_A = \omega_B$, i.e. A and B have the same natural numbers.

The omitting types theorem can be generalized as follows:

Theorem 8.9 Let T be a countable theory. For each i $\in \omega$, let pi be an essentially nonprincipal n_i-type over T. Then T has a model which omits pi for each i $\in \omega$.

The above generalization of the omitting types theorem to can be applied to prove the following:

Theorem 8.10 (Keisler and Morley). Let A be a countable model of ZF set theory. Then A has a proper elementary end extension, i.e. a proper elementary extension B such that rank(a) < rank(b) for all a $\in$ |A|, b $\in$ |B| \ |A|.

Definition 8.11 Let T be a complete theory. A model A of T is said to be prime if every model of T has an elementary submodel, which is isomorphic to A.

Examples are the theory of algebraic closed fields of characteristic 0. This theory has a prime model: the algebraic closure of the rational fields.

Similar example is T, the theory of algebraically closed fields of prime characteristic p. The prime model is the algebraic closure of Fp = (Fp, +, −, ·, 0, 1).

Generalizing let A be any field and let T = Algebraic closed fields ∪ {diagram of A}. T is complete because theory of algebraic closed fields admits elimination of quantifiers. The prime model of T is A, the algebraic closure of A.

T = Real Closed Ordered Fields. The prime model is Q = the real closure of the ordered field Q = (Q,+,−,·,0,1,<).

T = Real Closed Ordered Fields ∪ (diagram of F) where F is any ordered field. The prime model is F, the real closure of F.

An example of a complete countable theory with no prime model. Let T have 1-place relations Ri(x), i ∈ ω, and axioms saying that the Si's are independent, i.e. any nontrivial Boolean combination of them is nonempty. The completeness of T can be proved by quantifier elimination. It is easy to see that no model of T is prime. Remark. It is clear from the definition of prime model that any elementary submodel of a prime model is prime. Example 6 shows that a prime model may have proper elementary submodels.

The purpose of this section is to establish necessary and sufficient conditions for a complete theory T to have a prime model, and to establish the unique- ness of prime models when they exist. The following lemma is a restatement of the omitting types theorem in the special case when T and p are complete.

Lemma 8.1 Let T be a countable complete theory, and let p ∈ Φn(T) be a complete n-type over T. The following are equivalent.

(i)p is principal over T.
(ii) p is realized in every model of T.

Proof Suppose p is principal, e.g., generated by φ ∈ Fn(T). In particular T ∪ {∃v1 ···vn φ(v1,...,vn)} is consistent. Since we are assuming that T is complete, it follows that T ⊨ ∃v1 ···vn φ(v1,...,vn). Let A be any model of T. Since A ⊨ ∃v1 ···vn φ(v1,...,vn), there exist a1,...,an ∈ |A| such that A ⊨ φ(a1,...,an). Then clearly the n-tuple a1,...,an realizes p. Conversely, when p is nonprincipal over T. Since p is complete, it follows that p is essentially nonprincipal over T. Hence by the omitting types theorem, there exists a model of T which omits p.

Definition 8.12. A structure A is atomic if for every a1, . . . , an ∈ |A| the complete n-type realized by a1, . . . , an is principal over Th(A).

Theorem 8.11 (Vaught). Let T be a countable complete theory. The following are equivalent for A ∈ class of models of T.

(i) A is prime.
(ii) A is countable and atomic.

Proof (i) ⇒ (ii). Assume A is prime. Since T has a countable model B and A is isomorphic to an elementary submodel of B, it follows that A is countable. To show that A is atomic, let a1,...,an ∈ |A| and let p be the complete n-type over T realized by a1, . . . , an. Since A is prime, p is realized in every model of A. Hence by the previous lemma it follows that p is principal. Thus A is atomic.

(ii) ⇒ (i). Assume that A is countable and atomic. Let |A| = {an : n∈ω} be an enumeration of |A|. Let B be a model of T. Create an elementary embedding of A into B. Inductively have chosen b0,...,bn−1 ∈ |B| : (A,ai)i<n ≡ (B,bi)i<n. Since A is atomic, so is (A,ai)i<n. For, let c1,...,ck ∈ |A|. The complete (n + k)- type realized by ⟨a0, . . . , an−1, c1, . . . , ck⟩ is principal over T , say generated by φ(u0,...,un−1,w1,...,wk).

Let ψ(w1,...,wk) ≡ φ (a0, . . . , an−1, w1,...,wk). Then j generates the complete k-type realized by c1, . . . , ck over Th((A, ai)i<n). Let pn ∈ Φ_1((A,ai)i<n) be the complete 1-type realized by an. Since p_n is principal, it follows by the previous lemma that p_n is realized in (B,bi)i< n, say by bn ∈ |B|. Then (A, ai) i≤ n ≡ (B, bi) I ≤ n. Finally (A, ai) i∈ω ≡ (B, bi)i∈ω. Thus ai → bi gives an elementary embedding of A into B.

8.6 FILTERS AND FIELDS

Here we brief on prime ideals form the functorial model viewpoint considering the preceding chapters. When using the functor Ω to pass from **Top** to **Fram**, all set-theoretic elements of a space are lost, but – using a fundamental idea of category theory – one can as well work on the function spaces. Any "point" *px*: 1 → *X* in **Top** is mapped to a morphism Ω(*px*): Ω(*X*) → Ω(1). The open set lattice of the one-element topological space Ω(1) is just (isomorphic to) the two-element locale 2 = { 0, 1 } with 0 < 1. After these observations it might be reasonable to define the set of points of a locale *L* to be the set of frame morphisms from *L* to 2. However, there is no guarantee that every point of the locale $\Omega(X)$ is in one-to-one correspondence to a point of the topological space X,e.g., consider again the indiscrete topology, for which the open set lattice has only one "point".

Before defining the required topology on pt(X), it is worthwhile to clarify the concept of a point of a locale further. The perspective motivated above suggests to consider a point of a locale L as a frame morphism p from L to 2. But these morphisms are characterized equivalently by the inverse images of the two elements of 2. From the properties of frame morphisms, one can derive that $p-1(0)$ is a lower set, since p is monotone, which contains a greatest element $ap = \bigvee p-1(0)$ (since p preserves arbitrary suprema). In addition, the principal ideal $p-1(0)$ is a prime ideal since p preserves finite infima and thus the principal ap is a meet-prime element. Now the set-inverse of $p-1(0)$ given by $p-1(1)$ is a completely prime filter because $p-1(0)$ is a principal prime ideal. It turns out that all of these descriptions uniquely determine the initial frame morphism. We sum up:

A point of a locale L is equivalently described as:

a frame morphism from L to 2

a principal prime ideal of L

a meet-prime element of L

a completely prime filter of L.

All of these descriptions have their place within the theory and it is convenient to switch between them as needed.

Theorem 8.12 (Vaught). Let T be a countable complete theory. If T has a prime model, it is unique up to isomorphism.

Proof. (Exercises and the preceding theorem).

Theorem 8.13 (Vaught). Let T be a countable complete theory. The following are equivalent.

(i) T has a prime model.

(ii) T has an atomic model.

(iii) Every principal n-type over T is included in a complete principal n-type over T.

Remark. A Boolean algebra B is said to be atomic if for all b ∈ B, b/= 0, there exists a ≤ b such that a is an atom, i.e. a/= 0 and there is no a1 such that $0 < a1 < a$. The condition (iii) in the theorem can be restated as follows: for each $n \in \omega$, the Boolean algebra Bn(T) is atomic.

Corollary 8.2 (Vaught). Let T be a countable complete theory. Suppose Φn(T) is countable for each $n \in \omega$. Then T has a prime model.

Proof Assuming that Φn(T) is countable, can prove that Bn (T) is atomic, i.e., every principal n-type p over T is included in a complete principal n-type over T.

Definition 8.13 Let A,B be structures with $S_A = S_B$ and $|A| \subseteq |B|$. We say that A ⊆e B if for all formulas φ(x1,...,xn) with only the free variables shown, and all a1,...,an ∈ |A|, A ⊨ φ(a1,...,an) if and only if B ⊨ φ(a1, . . . , an). We then say that A is an elementary substructure of B, or B is an elementary extension of A.

Let A, B be as above. We say that A ⊆ B (A is a substructure of B, B is an extension of A) if the above condition holds for atomic formulas φ. Example. Algebra is full of examples of substructures. E.g. if A and B are groups, A ⊆ B if and only if A is a subgroup of B. Similarly for rings, fields, linear orderings, etc.

Example. Let Q = (Q,+,−,·,0,1,<) and R = (R,+,−,·,0,1,<). Then Q⊆R, but Q $\not\subseteq_e$ R.

Proposition 8.2 If A ⊆e B then A ≡ B.

Proposition 8.3. If A, B ∈ class of models of T where T admits elimination of quantifiers then A⊆B implies A⊆e B.

The fact that the theory of algebraically closed fields admits elimination of quantifiers yields the following,c.f.chapter 9:

Definition 8.14 (A. Robinson). A theory T is said to be model complete if for all A,B∈ class of models of T, A⊆B implies A⊆e B.

The above says that if T admits elimination of quantifiers, then T is model complete. The converse is not true, for example, T = Th(Z, +, −, 0, 1, <) = Presburger arithmetic.

However, the theory of algebraically closed fields is model complete and admits elimination of quantifiers. Important publications on these areas are, for example, Macintire (1977) on 20 years of p-adic model theory. The algebraic importance, for example, is that model completeness expresses the Hilbert Nullstellensatz.

Crossing to Chapters 4–7, to have concrete examples on the application areas from the authors and McIntires publications, we can get specific and define an algebraically closed groups from the authors ASL 1980s. For example, let the signature of groups be defined as above. Let R be the set of terms over the group signature. Define t1 ≡ t2 iff there is a condition p in , where is the F-generic filter UIC, such that p P– + t1=t2 ≡ is easily seen to be an equivalence relation by the definition of the forcing property. It is also a congruence relation with respect to the operations of the signature −1 and •, so that we have an instance of a generic filter. The congruence condition can be readily checked with respect to ≡. Now, let B be the canonical term algebra isomorphic to the quotient of R with respect to ≡.

Theorem 8.14 *B* = (B, •, −1, e) is an algebraically closed group.

Proof *B* is a group is obvious, because the defined congruence contains the group congruence. That it is algebraically closed is a direct consequence of theorem 6.7 and the foregoing discussion. In particular, every sentence consistent over group structures can be embedded or instantiated in a condition in the forcing property, thus it is satisfied by the generic model generated.

The result of the last two sections have some practical applications for solving word problems and completion of algebraic subtree replacement systems. The implications are that if one could construct a canonical term initial group for a group presentation, then one in effect has solved the word problem for groups, or any algebraic structure for that matter. Solving such problems is of great importance to computing.

The group problem has received considerable attention since the chapter of Knuth-Bendix (1970). Of course, that's not where things end. The problem of tree completion for algebraic subtree replacement systems is really what computing is interested is, c.f., Chapter 3. We can treat completing subtree replacement systems such that we can tree rewrite to models. Thus, carrying out symbolic computation. Here we take the cyclic group example, and show how to form an algebraically closed group with it. The group example, cyclic group of order six on the set {*x*, *y*}, is revisited here again: The normal forms are the six element of the set {e, *x*, y, xy, y_2, xy_2}.

Now consider the following three basic group axioms:

1.*e.x* = *x*

2.*x*–*1.x* = *e*

3.*(x.y). z* = *x. (y.z)*

Its completion thorough the Knuth-Bendix algorithm is attained by the following additional equations:

e–*1* = *e*

(y–*1)*–*1* = *y*

(y.y")–*1* = *y"1.y*

y–*1.(y.z)* = *z*

z.e= *z*

y. (y–*1.z)* = *z*

y.y–*1* = *e*

Let {*x*, *y*}* be the set of strings on the alphabet {*x*, *y*} such that $x_2=e$, $y_3=3$, and *xy*=*yx*, where *e* denotes the empty string (the identity of the structure). Every string in {a, K}* is equivalent to one of the six strings *e, a, K, ab*, K_2, ab_2. Thus these constitute a set of representatives for the above. Let {*e, a, K, ab*, K_2, ab_2} = G. the canonical term algebra is defined by putting the obvious group structure on the six-element set.

a.b = *ab*; *e.x* = *x* for all *z* in G; $b.\ b_2= e$; $ab.b_2 = x$; $ab.ab_2= e$; $a.b_2 = ab_2$; $b.ab_2= a$, etc.

In this case by picking a generating set for the diagram we can define a canonical initial model. The reader should not be dismayed by all that is found in the mathematics literature on unsolvability of such problems. For here, models are generated by infinitary conditions and sets, to present the model theory. We are not dealing with finitely generated models. They are models generated by infinitary forcing conditions. Neither we solve word problems in the proof theoretic sense.

8.7 FILTERS AND PRODUCTS

8.7.1 COMPLETE THEORIES

Recall from the preceding chapters:

Let I be a nonempty set. Let S(I) be the set of all subsets of I. A filter D over I is defined to be a set D < S(I) such that I ε D; if X, Y ε D, hen X ∩ Y ε D; if X ε D and X < Z < I, then Z ε D. Note that every filter D is a nonempty set since I ε D. For example, filters are the trivial filters D = {I}. The improper filter D = S(I). For each Y < I, the filter D = {X <I; Y < X}; this filter is called the principal filter generated by Y. D is said to be a proper filter iff it is not the improper filter S(I).

Let E be a subset of S(I). By the filter generated by E we mean the intersection D of all filters over I which include E: D = ∩ {F: E < F and F is a filer over I}. E is said to have the finite intersection property iff the intersection of any finite number if elements of E is nonempty.

Can prove that the filter D generated by E, any subset E of S(I), is a filter over I.

Certain products on filters called reduced products are what we will apply to fields on what follows to create model factorizations over algebraically closed fields. The specific techniques and application areas, modula Vaught's obvious accomplishments, are new as far as we are aware, especially in view of Section 6.

Suppose I is a nonempty set, D a proper filter over I, and for each i∈ I, Ai is nonempty. Let C = ∏i∈ I Ai be the cartesion product of the sets. C is the set of all functions f with domain I s.t. for each i∈ I, f(I) Ai. For functions f and g i∈ C, say that f and g are D-equivalent, f =D g, iff {i∈ I: f(i)=g(i)} ∈ D. The relation =D is easily proved an equivalence relation over C. Let fD. be the equivalence class of f. fD = {g ∈ C: f =D g}.

Reduced product for sets Ai modula D is defined to be the set of all equivalence classes of =D , dentoed by ∏D Ai. ∏D Ai.={fD: f∈∏I∈I Ai}.

On models reduced products are defined based on D, a proper filter over I, are defined as follows. For a language L, with usual conventions on functions and relation symbols, let Ri be a model for L. The reduced product ∏D Ri. Is the model for L as follow

(I) the universe set is $\prod D$ Ai.; (II) Let P be an n-placed relation symbol of L. The interpretation of P in $\prod$DRi.is the relation S s.t. S(f1D...

fnD) iff {i∈ I,Ri (f $^{1(i)}$ .... f$^{n(i))}$} ∈ D (III) Let F be an n-placed funciton symbol of L. The F is interpreted in $\prod$DRi.by the function H, H

(f1D....fnD) = <Gi (f1 (i) ...fn(i): i∈ I> D Let c be a constant of L. Then c is interpreted by th element b ∈ $\prod$D Ai, where b = <ai : i∈ I> D.

8.7.2 FRAGMENT CONSISTENT MODELS AND PRODUCT FIELDS

This section is a preliminary outline to how fragment consistent models can be applied to create models for algebraically closed fields from the authors 2005–2012 ASL-AMS briefs.

Starting with the unique factorization theorem on Galois fields. Consider polynomials f(D) over an algebraically closed Galois field of prime characteristic p. The above field is complete based on the stated theorems. Let us apply Vaught's theorem:.

Lemma 8.3 The model completion T* on the theory T of algebraic closed fields of characteristic (p=0 or prime) defines a proper filter on the set of T formulas.

Theorem 8.15 (Prime Model Factorization) Consider polynomials factorization over the algebraic closed fields of characteristic p (p=0 or prime) on polynomials definable by T fragments. Let M be the prime model to T. Then there are prime models Mi, modeling the factors, respectively, such that there is a reduced product based on a proper filter D, defined on T, principal omitting type on fragments with ΠD Mi monomorphically embedded in M.

Proof The proof applies the following:

(a) The theory of algebraic closed fields is a model completion on theory of fields. Let T* denote the former model completion.
(b) Compactness theorem (Section 2) and fragment consistency Theorem 5.2 are applied to have model completeness. c.f. Vaught theorems, for example, Theorem (Vaught)

Let *T* be a countable complete theory. The following are equivalent.

(i) *T* has a prime model.

(ii) *T* has an atomic model.

(iii) Every principal *n-type* over *T* is included in a complete principal *n-type* over *T.*

Corollary 8.3 There is a reduced product at Theorem 6.2 each model Mi omit n-types split on the polynomial factors such that one and only one specific polynomial factor is modeled by each model Mi.

The model completion filter is not necessarily strong enough to carry a direct split on the field models. Nourani (2006) is on example generic filters that might be applied to have further specifics.

Theorem 8.16 Ri is a prime model of T iff Ri are elementarily embedded in every countable model of the fragment of T* that Ri models.

Corollary 8.3 There is a reduced product at Theorem 8.3 each model Mi omit n-types split on the polynomial factors such that one and only one specific polynomial factor is modeled by each model Mi.

The model completion filter is not necessarily strong enough to carry a direct split on the field models. Nourani (2006) is on example generic filters that might be applied to have further specifics.

Theorem 8.17 Ri is a prime model of T iff Ri are elementarily embedded in every countable model of the fragment of T* that Ri models.

Proposition 8.4 Every prime model for a complete countable theory has a creative certificate Ω.

Proof follows from the following: A newer look at models and extensions since the author's two decades on initial models a constructive signature Ω for algebras of a class C is a set that implicitly determines the remaining operations. For example the successor operation is primitive to natural numbers, and either · or + is primitive for Boolean algebras. From these observations we can define field extension as follows. We call Ω a creative certificate. By definition a model A for T is prime iff every model of T has an elementary submodel that is isomorphic to A. From Vaught assuming complete countable theory T, if T has a prime model it is unique up to isomorphism. The first author had proposed the following on an abstract note (June, 2005) further specifics are coannounced on an abstract at AMS, May 2008.

Lemma 8.4 The generic diagram defines a creative certificate in the sense that the model is definable by a certificate signature.

Proof Since there is a subsignature that Ω defining the generic diagram uniquely determines the model for T, by Vaught's theorem the prime mode must be unique up to isomorphism, every model for T has an elementary embedding to A.

Lemma 8.5 The generic diagram defines a succinct certificate for prime models. The certificate functions define representative for the equivalence classes that create a model.

EXERCISE

1. Prove the following proposition.
   (i) Every uncountable algebraically closed field is saturated.
   (ii) Q and Fp are not saturated.
   (iii) There exist countable saturated algebraically closed fields of every characteristic.
   Hint: Proof. A complete 1-type over X is virtually the same thing as a simple extension of the subfield A generated by X. Apply simple field extension.
2. Prove Vaught's test Let T be a theory, which (i) has no finite models; (ii) is κ-categorical for some $\kappa \geq \text{card}(T)$. Then T is complete.
   Hint: Apply Lowenheim-Skolem-Tarski theorem.
3. Prove that the theory of dense linear orderings without end points is complete.
4 With R = (R, +, −, ·, 0, 1, <) = the ordered field of real numbers. Z = (Z, +, −, ·, 0, 1, <) = the ordered field of rationals. Show that T theory of dense linear ordering without end points has a prime model: (Q, <).

4. The theory of differential fields, DF, has the following set of axioms: (a) field axioms (b) $\forall x\, \forall y\, (x + y)^\wedge = x^\wedge + y^\wedge$ (c) $\forall x\, \forall y\, (x \cdot y)^\wedge = x^\wedge \cdot y + x \cdot y^\wedge$
   Consider $^\wedge$ is a 1-place operation symbol. A differential field is a model of these axioms. Let A be any field. The polynomial ring A[x] can be made into a differential ring by interpreting $'$ as the formal derivative, that is, if f = anxn +···+a1x+a0, ai ∈|A|, then f′ =nanxn−1 +···+a1.
   Prove that every differential field, which is finitely generated over the relational field is isomorphic to a differential subfield of N.
5. Definition. Let F be an ordered field. A real closure of F is an ordered field G ⊇ F so that (1) G is algebraic over F; (2) G is real closed.
   Prove that any ordered field F has a real closure.
   Hint: Apply Zorn's lemma.
6. Theorem Let T be a countable theory. For each i ∈ ω, let pi be an essentially nonprincipal ni-type over T. Then T has a model which omits pi for each i ∈ ω.
   Hint: Generalization of the proof of the omitting types theorem.
7. (Robinson). If T is model complete then T is a Π2-theory.
8. Prove the basic omitting types theorem using Henkin proof of the completeness theorem.
   Hint: Enumrate enumeration of all formulas with no free variables other than v, such that signature (φ(v)) ⊆ sigarue of T∪ with enriched Henkin constants.
9. Let us say that φ, ψ ∈ Fn (T) are equivalent over T if
   $T \models \forall v1 \cdots \forall vn\, (\varphi \leftrightarrow \psi)$. The equivalence classes form a Boolean algebra Bn(T) where the Boolean operations are given by $[\varphi]\cdot[\psi] = [\varphi \wedge \psi]$
   $[\varphi] + [\psi] = [\varphi \vee \psi]$ $-[\varphi] = [\neg\varphi]$ 1=[v1 =v1],0=[v1/=v1]
   (For n = 0 this Boolean algebra is sometimes known as the Lindenbaum algebra of T.)

Prove that a complete n-type over T is essentially the same thing as an ultrafilter on Bn(T).

Sn(T) is the Stone space of Bn(T).

10.Prove theorem 8.11

Hint: Back-and-forth techniques.

KEYWORDS

- **Henkin style**
- **Knuth-Bendix algorithm**
- **Prime models**
- **subtree replacement system**

CHAPTER 9

FILTERS AND ULTRAPRODUCTS ON PROJECTIVE SETS

CONTENTS

Functorial generic filters are presented and the applications to fragment consistency and algebraic products are briefed. Horn products are specific fragment consistent model examples treated. Specific topos project set models are based on ultrafilters and saturated models are presented. Some basics and definitions are briefs from preceding chapters, stated here for self-containment.

9.1 GENERAL DEFINITIONS

A non-empty subset F of a partially ordered set $(P,\leq)$ is a **filter** if the following conditions hold:

For every x, y in F, there is some element z in F, such that $z \leq x$ and $z \leq y$. (F is a **filter base**)

For every x in F and y in P, $x \leq y$ implies that y is in F. (F is an *upper set*) A filter is **proper** if it is not equal to the whole set P. This is often taken as part of the definition of a filter.

While the above definition is the most general way to define a filter for arbitrary posets, it was originally defined for lattices only. In this case, the above definition can be characterized by the following equivalent statement: A non-empty subset F of a lattice $(P,\leq)$ is a filter, if and only if it is an upper set that is closed under finite meets,i.e., for all x, y in F, we find that $x \wedge y$ is also in F.

The smallest filter that contains a given element p is a **principal filter** and p is a **principal element** in this situation. The principal filter for p is just given by the set $\{x \text{ in } P \mid p \leq x\}$ and is denoted by prefixing p with an upward arrow: .

The dual notion of a filter, i.e. the concept obtained by reversing all $\leq$ and exchanging $\wedge$ with $\vee$, is **ideal**.

A special case of a filter is a filter defined on a set. Given a set S, a partial ordering $\subseteq$ can be defined on the powerset **P**(S) by subset inclusion, turning (**P**(S),$\subseteq$) into a lattice. Define a **filter** F on S as a subset of **P**(S) with the following properties:

S is

1. S is in F. (*F is non-empty*)
2. The empty set is not in F. (*F is proper*)

3. If A and B are in F, then so is their intersection. (*F is closed under finite meets*)
4. If A is in F and A is a subset of B, then B is in F, for all subsets B of S. (*F is an upper set*) .

The first three properties imply that a **filter on a set** has the finite intersection property. Note that with this definition, a filter on a set is indeed a filter; in fact, it is a proper filter. Because of this, sometimes this is called a **proper filter on a set**; however, as long as the set context is clear, the shorter name is sufficient.

A **filter base** (or **filter basis**) is a subset B of $\mathbf{P}(S)$ with the following properties:

1. The intersection of any two sets of B contains a set of B
2. B is nonempty and the empty set is not in B

Given a filter base B, one may obtain a (proper) filter by including all sets of $\mathbf{P}(S)$ that contain a set of B. The resulting filter is said to be generated by or spanned by filter base B. Every filter is *a fortiori* a filter base, so the process of passing from filter base to filter may be viewed as a sort of completion.

If B and C are two filter bases on S, one says C is **finer** than B (or that C is a **refinement** of B) if for each $B_0 \in B$, there is a $C_0 \in C$ such that $C_0 \subseteq B_0$.

For filter bases A, B, and C, if A is finer than B and B is finer than C then A is finer than C. Thus the refinement relation is a preorder on the set of filter bases, and the passage from filter base to filter is an instance of passing from a preordering to the associated partial ordering.

Given a subset T of $\mathbf{P}(S)$ we can ask whether there exists a smallest filter F containing T. Such a filter exists if and only if the finite intersection of subsets of T is nonempty. We call T a **subbase** of F and say F is **generated** by T. F can be constructed by taking all finite intersections of T which is then filter base for F.

Let S be a nonempty set and C be a nonempty subset. Then $\{C\}$ is a filter base. The filter it generates (i.e., the collection of all subsets containing C) is called the **principal filter** generated by C.

A filter is said to be a **free filter** if the intersection of all of its members is empty. A principal filter is not free. Since the intersection of any finite number of members of a filter is also a member, no filter on a finite set is free, and indeed is the principal filter generated by the common intersection of all of its members. A nonprincipal filter on an infinite set is not necessarily free. The Fréchet filter on an infinite set S is the set of all subsets of S that have finite complement. The Fréchet filter is free, and it is contained in every free filter on S.

A uniform structure on a set X is (in particular) a filter on $X \times X$.

A filter in a poset can be created using the Rasiowa-Sikorski lemma (c.f. chapter10) and Nourani (2007), often used in forcing.

The set is called a *filter base of tails* of the sequence of natural numbers. A filter base of tails can be made of any net using the construction. Therefore, all nets generate a filter base (and therefore a filter). Since all sequences are nets, this holds for sequences as well. During 2005–2007 the author explored positive and Horn fragment categories based on his 1981 on positive forcing on Kiesler fragments.

Positive forcing had defined T^* on a theory T to be T augmented with inductive definability schemas on the generic diagram functions, for a model M for T. The author defined positive local reliability on 2005–2007 publications (ASL). Generic filter models, filters on Horm models, and saturation on fragment consistent models are applied towards projective set models and resembling what the author terms certain *projective determinants* (University of California Mathematics notes Note 2012).

9.2 GENERIC FUNCTORS AND LANGUAGE STRING MODELS

Creating limits amounts to defining generic sets on $L_{w1,w}$ for $\mathbf{L}_{\mathbf{w1, K}}$. The present theory creates limits for Infinitary languages to define generic models with language categories. Generic sets for infinitary languages have been applied by this author to define Positive Forcing on Toposes.

Define a functor F: $\mathbf{L}_{\mathbf{w1, K}} \rightarrow$ **Set** by a list of sets Mn and functions fn. Let us refer to the above functor by the **generic model functor** since it defines generic sets from language strings to form limits and models. The model theoretic properties are not defined in the present chapter and are to be presented elsewhere by this author. The proof for the following theorem follows from the sort of techniques applied by the present author to define generic sets with the fragment $\mathbf{L}_{\mathbf{w1, K}}$.

Definition 9.1 Let $\Im$= (S, ≤, f) be a positive forcing property, we say that a subset G of S is positive generic iff

(i) p in G and q < p implies q in G;
(ii) p, q in G, implies there is an r in G such that p ≤ r and q ≤ r;
(iii) for each sentence j, there exists p in G, such that either p ||– + φ or there is no q in S, q >p such that q ||—+ φ.

A special case of the above definition is when S consists of sets of formulas. For such S, we can make a substitution: subset relation for <. Clause (iii) can then be stated:

(iii)' for each sentence φ, there exists p in G such that either p ||– + φ or p U {φ} is not a condition for *i.*

Definition 9.2 Let (M, a)*c* in C be defined such that M is a model for L and each constant *c* in C has the interpretation a in M. the mapping *c* → ac is an assignment of C in M. We say that (M, a)*c* in C is canonical model for a presentation P iff the assignment *c* → a maps C onto M, that is, M=(a:*c* in C).

Definition 9.3 A generic set is said to **generate** *M* iff *M* is a canonical model and every sentence φ of L[C], which is forced by some p in G holds in *M.*

Definition 9.4 M is a generic model for a condition p iff M is generated by some generic set G, which contains p.

Definition 9.5 A theory is said to be inductive if T is equivalent to a set of ∀∃ sentences- sentences of the form ∀x1… ∀xn∃ x1..∃ xn *y*, where *y* is quantifier-free.

The function set might be S1 Skolem functions for the set theory example.

The following theorem presented itself in the course of writing the proofs for the present chapter and is stated. It is a consequence of the positive forcing basis since the past decade.

Theorem 9.1 The power set of an inductive theory defined on $L_{w1,k}$. generates a generic model for the theory, provided the inductive theory has a defined generic diagram for its initial model.

Proof If the inductive theory has a definable G-diagram for its initial model M, then its powerset has the inductive closure as a maximal member. Since M is a canonical model, for every sentence φ forced by some p in the powerset, if φ is atomic, then by definition it must be in p to be forced. Thus it is modeled by M. For an arbitrary formula φ, if j is forced by some p, then there is an r >=p such that r forces φ(l), where φ(l) is obtained from φ by a Skolemization by G-diagram terms for the existential quantifiers. Therefore, φ ε r, M models φ, and M is a generic model. –

Theorem 9.2 The Generic model functor defines a positive generic model.

Proof The limit M from the arrow diagram in theorems from Chapter 3 (c.f. Theorem 9.3) is a positive generic model for conditions definable on $L_{w1,\,k.}$ For M is generated by positive generic sets definable on $L_{w1,\,k}$.

These are the theorems and Chapter 3 that allows us to create a limit model on the topos vertex for projections and presented here for self-containment.

Theorem 9.3 The generic model functor has a limit.

Proof Chapters 3–5.

Let us create a limit for the functor F: $\mathbf{L}_{\mathbf{w1,\,k}}$ → **Set** in **Set** as follows. The functor F is a list of sets Fn, consisting of:

(a) the sets corresponding to a free structure on Lw1, B, for example, to f(t1, t2,.. tn) in $L_{w1,\,k}$. there corresponds the equality relation f(t1, …, tn)=ft1…tn in Set;
(b) the functions fi:Fi+1 → Fi. Form the product set PiFi. Let l = {l0, l1, l2, ….}, with each ln ∈ Fn, and its projections pn: PiFi → Fn, forming the following diagram.

```
F0  ←   F1 ←               Fn ← Fn+1 ←
↑                                ↑
ΠiFi   ←   ←    include  ← M = Lim F
```

To form a limit at a vertex for commuting with the projections take the subset S of the strings in S, which match under f, that is, f sn+1 = sn, for all n. The above A-diagram is well-defined because by definitions above and theorem 9.1, sets which are generic with respect to a base set, for example the intended limit M are definable.

9.3 FUNCTORIAL FRAGMENT CONSISTENCY

From Chapter 4 we state a brief on the basics to carry on.

Goal is to create models for languages from fragments by an infinite limit. There are two ways to view it.

(a) Take $L_{w1,\,k}$ language fragments, define w-chain models, and back and forth to a limit diagram model.

(b) Define models for the Fi from elementary diagrams, that is, define a limit model by embedding from a D<A, G> model.

What complete theory can we fall onto? It has to be the theory Th(F), where F is the generic model functor defined to Set, that is,

Th (F: $L^{Op}_{\omega 1,\,k}$ → **Set**).

Starting with $L_{\omega 1,\,k}$ language fragments, define w-chain models, and back and forth to a limit model. Remember we had defined the fragment by letting C be a countable set of new constants symbols and form the first order language K by adding language fragments, define w-chain models, and back and forth to a limit model.

Let $L_{\omega 1,\,k}$ be the least fragment of $K_{\omega 1,\,w}$ which contains L<A>. Each formula j in K<A> contains only finitely many *c* in C. This implies when raking leaves on the trees, there are only finite number of named branches claimed by constant names. The infinite trees are defined by function names, however. The functions define the model with the constants.

Lemma 9.1 For a chain Aβ, β < α, of models, ∪ Aβ is the unique model with universe ∪ Aβ which contains each Aβ as a submodel.

Proof c.f. preceding chapters.

A functor **F**: $L^{Op}_{\omega 1,\,k}$ → **Set** can be defined by sets **Fi**, where the Fi's are defining a free structure on some subfragment of $L_{\omega 1,\,\mathbf{K}}$. To be specific we can define the subfragment models A(Fi) straight from the w-inductive definition of the Infinitary fragment. F0 assigns names to the Set members, for example, F1 can define 1-place functions and relations, so on and so forth.

Taking the languages defined above and the Robinson's consistency theorem define Functorial Limit Chain models as follows. We shall refer to it by FLC-models (Nourani, 1996).

Let A and B, be models for Fi and Fi+1, respectively. Let A ≡L B, and f: A <L B mean the L-reduct of A and B are elementarily equivalent and that f is an elementary embedding of A|L into B|L. Let A *FLC* model be the limit model defined by the elementary chain on the L-reducts of the models defined by the Fi's. A specific *FLC* model is defined by theorem 9.5′s proof.

Theorem 9.4 There is an elementary chain *FLC* model for L, where L is $L_{\omega 1, k.}$

Proof Chapter 4.

9.4 FILTERS

Let I be a nonempty set. Let S(I) be the set of all subsets of I. A filter D over I is defined to be a set D < S(I) such that I ε D; if X, Y ε D, hen X ∩ Y ε D,; if X ε D and X < Z < I, then Z ε D. Note that every filter D is a nonempty set since I ε D, example filters are the trivial filter D = {I}. The improper filter D = S(I). For each Y < I, the filter D = {X <I; Y < X}; this filter is called the principal filter generated by Y. D is said to be a proper filter iff it is not the improper filter S(I). Let E be a subset of S(I). By the filter generated by E we mean the intersection D of all filters over I which include E: D = ∩ {F: E < F and F is a filer over I}. E is said to have the finite intersection property iff the intersection of any finite number if elements of E is nonempty.Can prove that the filter D generated by E, any subset E of S(I), is a filter over I.

Definition 9.6 Let Q be a poset, F a family of sets, G subset of Q. The G is F-generic iff the following holds,

(i) whenever p in G and q < p, then q is in G.

(ii) whenever p, q in G, then there exists an r in G with r> p and r > q. In particular, any two elements of G are compatible.

(iii) $G \cap D = \varnothing$ for any dense D subset of Q with D in F.

Note that a subset D of Q is said to be dense iff for every p in Q there is a q in D with p < q.

Theorem 9.5 $\wp$(T*) is generating a generic model with the F-generic filter.

Corollary 9.1 The positive forcing T* is a F-generic filter.

Proof(s) Let F be a family of functions, which minimally defines the universe with the generic diagram for M. From Chapter 4 and above theorems the inductive closure T* (P) generates a generic model, with T#(P) = T(P) + (I[f]: f F) is defined. For algebraic structures the universal reduct of T# (P) is sufficient. We refer to the universal reduct by T* (P). That reduct instantiates the forcing property where $\wp$(T* (P)) is positively generic for T(P). From Theorems 9.1 and 9.2 and the preceding we have exhibited an explicit construction for an F- generic filer set where there is a dense filer base (c.f. ASL, 2006).

9.5 STRUCTURE PRODUCTS

The following propositions 9.1 to 9.4 are from basic model theory, c.f. Kiesler, (1977). We use the notation |A| for the universe of A and ||A|| for the cadinality of A.

Proposition 9.1 Let φ be a universal sentence. Then φ is a (finite) direct product sentence iff φ is equivalent to a universal Horn sentence.

Proposition 9.2 Let φ be an existential sentence. Then φ is a (finite) direct product sentence iff φ is equivalent to existential Horn sentence.

Proposition 9.3 Let φ be a universal sentence. Then φ is a (finite) direct product sentence if and only if φ is equivalent to a universal Horn sentence.

Proposition 9.4 Let φ be an existential sentence. Then φ is a (finite) direct product sentence if φ is equivalent to an existential Horn sentence.

Now let us apply generic filters on positive forcing to have a closer reach to Horn models.

Proposition 9.5 Let I be the set T*. Let φ (x1…xn) be a Horn formula and let 𝔑iiεI be models for language L. let a1….an ε∏iεI Ai. The 𝔑i are fragment Horn models. If {iεI: 𝔑i ⊨ φ[(a1(i) …an(i)]} the direct product over D on 𝔑i ⊨ φ[a1D …anD], where D is the generic filter on T*.

Proof Positive forcing had defined T* to be a theory T on $\mathbf{L}_{w1, K}$ augmented with induction on the generic diagram functions. Based on the infinitary counterpart to the Robinson's consistency functorial Fragment Limit Chain models, were defined(1996) proving there are elementary chain FLC model for L, $\mathbf{L}_{w1, K.}$

9.6 COMPLETING THEORIES AND FRAGMENTS

Consider a complete theory T in L. A formula φ (x1,., xn) is said to be complete (in T) iff for every formula ψ(x1,., xn), exactly on of T models φ → ψ or T models φ → not ψ olds. A formula θ (x1, …, xn) is said to be completable (in T) iff there is a complete formula φ (x1,., xn) with T models φ → θ. If that can't be done θ is said to be incompletable.

A theory is aid to be atomic iff every formula of L which is consistent with T is completable in T. A model 𝔑 is said to be atomic iff every n-tuple a1, …, an in A satisfies a complete formula in Th(𝔑).

A mapping f:A →B is an elementary embedding of 𝔑 into B in symbols written 𝔑 < B iff for all formauls φ (x1,., xn) of L and n-tuples a1, …, an in A, we have Â models φ [a1…an] iff B models φ (fa1…fan).

An elementary embedding of Â into B is thus the same as an isomorphism of 𝔑 into B.

Let us brief on what preceding bases there are on model-theoretic forcing, apart from positive forcing and Kiesler's elegant omitting types. Let S be a class of similar structures with signatures signature S.

Definition 9.7 A formula X is defined in structures M with signatures S iff all constants, functions and relation symbols of X are interpreted in M. M ⊨ and M ||– coincide except on ¬*x* where M ||– *x* for *x* ¬ *y* iff there is no extension M' of M, such that M' $||=$ *y*. If Σ includes a single structure M, then M $||-$ *x* iff M $\models x$, since there is no M' in S that extends M. Let K be a nonempty consistent set of sentences. Let S_K be the class of all structures consistent with K, that is, which are sunstructures of models of K, together with the empty set. S_K is K-generic iff all the sentences of K are defined in M and for every *x* defined in M, M $||-$ either *x* or $\neg$ ***x***. The following accomplishments on this section before Vaught's prime model theorems are due in part to Robinson et. al times. and Kiesler.

Theorem 9.7 Let M $\in$ S. Then M is contained in a K-generic structure M* $\in$ S.

Proof Applies Lindenbaum's lemma. (Exercises.)

Alternative definition of K-genericity is K is defined in M and forcing and satisfaction coincide. Let K ^{F}(M) be the set of all sentences defined in M and are weakly forced by M: K $^{F=}$K ^{F}(ϕ) is called the forcing companion of K.

Theorem 9.8 In order that a structure M $\in$ S be K-generic it is necessary and sufficient

(i) that sentences of K be defined in M;
(ii) and that for every sentence *x* defined in M, M $\models x$ iff M $||-$ *x*.

Proof One direction is direct and the second by induction on complexity of formulas.

Theorem 9.9 Let M and M' be K-generic structures in $\Sigma_{\text{K such that}}$ M' $\supset$ M. Then M's is an elementary extension of M. (Exercises).

Definition 9.8 A theory T is said to be inductive if T is equivalent to a set of $\forall x\ldots$"$\exists y\ldots\exists\varphi$ sentences.

Proposition 9.6 If T is inductive, then the union of any increasing chain $M_0 \subseteq M_1 \subseteq M_2$ ..... of models of T is a model of T.

Theorem 9.10 The class of all K-generic structures is inductive, that is, is union of chains closed.

Let $K_\forall$ be the set of sentences, which are defined in and deducible from K: the universal part of K. Two sets of sentences with the same universal part are considered universally equivalent. A structure is consistent with K iff it is consistent with $K_{\forall.}$

Theorem 9.11 Let M be K-generic structure and let M' ⊃ M be a model of K. Let X be any ∀∃ sentence defined in M'. The X holds in M.

Proof Exercise 9: Robinson.

9.7 PRIME MODELS AND MODEL COMPLETION

9.7.1 COMPLETING THEORIES AND FRAGMENTS

Consider a complete theory T in L. A formula φ (x1,., xn) is said to be complete (in T) iff for every formula ψ(x1,., xn), exactly on of T models φ → ψ or T models φ→ not ψ olds. A formula θ (x1, …, xn) is said to be completable (in T) iff there is a complete formula φ (x1,., xn) with T models φ → θ. If that can't be done θ is said to be incompletable.

A theory is said to be atomic iff every formula of L which is consistent with T is completable in T. A model 𝔑 is said to be atomic iff every n-tuple a1, …, an in A satisfies a complete formula in Th(𝔑).

A mapping f:A → B is an elementary embedding of 𝔑 into B in symbols written 𝔑 < B iff for all formulas φ (x1,., xn) of L and n-tuples a1, …, an in A, we have 𝔑 models φ [a1…an] iff B models φ (fa1…fan). An elementary embedding of 𝔑 into B is thus the same as an isomorphism of 𝔑 into B. Consider a complete theory T in L. A formula φ (x1,., xn) is said to be complete (in T) iff for every formula ψ(x1,., xn), exactly on of T models φ → ψ or T models φ →,not ψ olds. A formula θ (x1, …, xn) is said to be completable (in T) iff there is a complete formula φ (x1,., xn) with T models φ →, θ. If that can't be done θ is said to be incompletable.

A theory is aid to be atomic iff every formula of L which is consistent with T is completable in T. A model 𝔑 is said to be atomic iff every n-tuple a1, …, an in A satisfies a complete formula in Th(𝔑). Lyndon's preservation theorem for positive sentences. A formula is said to be positive if it is built up from atomic formulas using only ∀, ∃, ∧, ∨ (and not using ¬, →, ↔). A positive theory is a theory with a set of positive sentences as axioms.

Theorem 9.12 (Lyndon) Let T be a theory, which is preserved under homomorphic images, that is, any homomorphic image of a model of T is a model of T. Then T is a positive theory (the converse is easy).

For example, the theory of groups is positive, and each homomorphic image of a group is a group. The theory of commutative rings is not positive because of the presence of the axiom 0/= 1. If we drop this axiom, the theory becomes positive, and every homomorphic image of a commutative ring is a commutative ring.

Definition 9.9 Let T and $T^{\#}$ be L-theories. $T^{\#}$ is a model-completion of T iff:

(a) T and $T^{\#}$ are mutually model consistent, that is, every model of T is embeddable in a model of $T^{\#}$ and vice versa;
(b) $T^{\#}$ is model-complete: 𝔜 ⊨$T^{\#}$ and B ⊨$T^{\#:}$ furthermore $\mathfrak{R} \subseteq$ B implies $\mathfrak{R}$ is an elementary substructure of B.
(c) If M ⊨T then $T^{\#} \cup$ Diagram (M) is complete.

When (a) and (*b*) hold for an L-theories T and T^, T^ is called a model-companion of T. For example, T theory of fields, $T^{\#,}$ the theory of algebraic closed fields is a model completion of T.

A mapping f: A →B is an elementary embedding of $\mathfrak{R}$ into B in symbols written $\mathfrak{R}$ < B iff for all formulas φ (x1,., xn) of L and n-tuples a1, …, an in A, we have $\mathfrak{R}$ models φ [a1…an] iff B models φ (fa1…fan). An elementary embedding of into B is thus the same as an isomorphism of $\mathfrak{R}$ into B.

Definition 9.10 (Language diagrams) From the first order language perspective the diagram language of $\mathfrak{R}$ is the expansion $L_{\hat{A}}$ formed by adding new constant symbols $c_{M,\ for}$ each element m ∈ $\mathfrak{R}$. The diagram expansion $\mathfrak{R}$ is the structure $\mathfrak{R}_M$ for $L_{\mathfrak{R}}$ such that each $c_{\mathfrak{R}}$ is interpreted by $\mathfrak{R}$.

The basic diagram of $\mathfrak{R}$ is the set D($\mathfrak{R}$) of all atomic and negated atomic sentences of $L_{\mathfrak{R}}$ which are true in $\mathfrak{R}_{M.}$ The elementary diagram of $\mathfrak{R}$ is the complete theory Th($\mathfrak{R}$M), that is, the set of all sentences of $L_{\mathfrak{R}}$ true in $\mathfrak{R}$. On function terms might be illuminating to notice that given a subset X ⊆ M, the expansion L_x and structure $\mathfrak{N}_x$ are defined in the natural way: if f: X $\rightarrow$ M, then $\mathfrak{N}_{fx}$ is the structure if L_x in which c_x is interpreted by f(x). The above is an indirect manner for reaching the basic diagram definitions in chapters 3 and 4, to have a closer to language reach. We conclude this section on generic structures with the following theorem from known model theory.

Theorem 9.13 A theory is preserved under substructures iff is is a universal theory.

Proof Let K be the calss of all substructures of models of T. Let $T^{\forall}$ be the set of all universal consequences of T. Can prove that K is the class of all models of T (Details on Exercises).

Theorem 9.14 (Robinson's test). A theory T is model complet iff for any A, B $\in$ Models (T) and monomorphism A $\rightarrow$ B, we can find an elementary extension A$*$ of A and a monomorphism B $\rightarrow$ A$*$ so that the diagram

B

↑ \

A⊆e A∗

Commutes.

9.7.2 PRIME MODELS

Let us have a glimpse on prime models here again since preceding chapters with respect to generic filters and saturation considerations. $\mathfrak{N}$ is said to be a prime model iff $\mathfrak{N}$ is elementarily embedded in every countable model of Th($\mathfrak{N}$).

Theorem 9.15 (Vaught)

The Following are equivalent:

(i) $\mathfrak{N}$ is a countable atomic model.

(ii) $\mathfrak{R}$ is a prime model.
(iii) $\mathfrak{R}$ is a countable prime model.

Lemma 9.2 T* is a principal proper filter.

Theorem 9.16 Ri is a prime model iff Ri are elementarily embedded in every countable model of the fragment of T* that Ri models.

Proof Author ASL Montreal, (2007), and Chapters 4 and 5.

9.7.3 GENERIC PRODUCTS

Definition 9.11 A formula is said to be positive iff it is built from atomic formulas using only the connectives &, v and the quantifiers ∀, ∃.

Definition 9.12 A formula φ (x1, x2, …, xn) is preserved under homomorphisms iff for any homomorphisms f of a model A onto a model B and all a1, …, an in A if A |= [a1, …, an] B |= [fa1, …, fan].

Theorem 9.10 A consistent theory is preserved under homomorphisms iff T has a set of positive axioms.

Proposition 9.7 Let I be the set T*. Let φ (x1…xn) be a Horn formula and let $\mathfrak{R}$I, i ∈ I be models for language L. Let a1….an ε∏iεI Ai. The $\mathfrak{R}$i are fragment Horn models.

If {i ∈ I: $\mathfrak{R}$i |= φ[(a1(i) …an(i)]} the direct product over D on $\mathfrak{R}$i |= φ[a1D …anD], where D is the generic filter on T*.

Theorem 9.11 Assume the continuum hypothesis $2^{\omega} = \omega^{+}$, then φ is completable in T* iff φ is equivalent to a universal Horn sentence.

Corollary 9.2 Theorem 9.11 can be carried on without CH when there is, for example, a specific Horn density on the fragments.

Theorem 9.12 $\mathfrak{R}$i is a prime model iff$\mathfrak{R}$i are elementarily embedded in every countable model of the fragment of T* that $\mathfrak{R}$i models

9.7.4 FILTERS AND ULTRAPRODUCTS ON PROJECTIVE SETS

Definition 9.13 Let κ be a finite or infinite cardinal number and M a model in some first-order language. Then M is called ***κ*-saturated** if for all subsets A of cardinality less than κ, M realizes all complete types over A. The model M is called saturated if it is |M|-saturated where |M| denotes the cardinality of M. That is, it realizes all complete types over sets of parameters of size less than |M|.

Countably varies from author groups: According to some a model *M* is called **countably saturated** if it is $\aleph_{1-}$saturated; that is, it realizes all complete types over countable sets of parameters. According to others, it is countably saturated if it is $\aleph_{0-}$saturated; that is, realizes all complete types over finite parameter sets.

The more intuitive notion – that all complete types of the language are realized is named **weak saturation**, which is the same as 1-saturation. The difference is that many structures contain elements, which are not definable, for example, there are **R**-elements, reals, not definable in the field language. However, there must be types to describe relationships with them. Thus there are sets of parameters from the structure the definition of types. For example, a *specific* increasing sequence having a bound can be expressed as realizing the type $\{x > Sek_n: n \in \omega\}$, using countably many parameters. If the sequence is not definable, this fact about the structure cannot be described using the base language, so an ω-saturated structure will have to be there.

Example saturated models:

(**Q**, <) – the set of rational numbers with their usual ordering is saturated. Intuitively, this is because any type consistent with the theory is implied by the order type; that is, the order the variables appear at tells you everything there is to know about their role in the structure. (**R**, <) – the set of real numbers with their usual ordering is *not* saturated. For example, take the type with a variable *x* containing the formula $x > -1/n$ for every natural number *n*, and the formula $x < 0$.

This type uses ω different parameters from **R**. Every finite subset of the type is realized on **R** by some real *x*, so by compactness it is consistent with the structure, but it is not realized, as it would imply an upper bound to the sequence $-1/n$, which is less than 0 (its least upper bound). Thus (**R**, <) is *not* ω_1-saturated, and not saturated. However, it *is* ω-saturated, for essentially the same reason as **Q** – every finite type is given by the order type, which if consistent, is always realized, because of the density of the order.

Both of these theories can be shown to be ω-categorical through the back-and-forth method. This can be generalized as follows: the unique model of cardinality κ of a countable κ-categorical theory is saturated.

9.7.5 SATURATION AND PRESERVATION THEOREMS

Given a theory T and a nonnegative integer n, let Φn(T) be the set of all formulas φ(v1, . . . , vn) with no free variables other than v1, . . . , vn, such that $\Sigma_\varphi \subseteq \Sigma_T$. We say that Y ⊆ Φn (T) is consistent over T if there exist A ∈ class of models for T, and a1,...,an ∈|A| such that A ⊨ φ(a1,...,an) for all φ∈Y. Note that by compactness, if each finite subset of Y is consistent over T then so is Y .

A realization of Y ⊆ Φn (T) is an n-tuple a1,...,an ∈ |A|, A ∈ class of models for T, such that A ⊨ φ(a1,...,an) for all φ ∈ Y . We then say that Y is realized in the model A.

We say that ψ ∈ Φn (T) is a logical consequence of Y ⊆ Φn (T) over T if every realization of Y is a realization of ψ.

Definition 9.14 An n-type over T is a set p ⊆ Φn (T) which is consistent over T and closed under logical consequence over T.

Definition 9.15 An n-type p over T is said to be complete if, for all φ ∈ Φn (T), either φ ∈ p or ¬φ ∈ p. The set of all complete n-types over T is denoted Ω_n(T).

Remark. Let us say that φ, ψ ∈ Φn (T) are equivalent over T if T |= ∀ v 1 · · · ∀ v n (φ ↔ ψ) . The equivalence classes form a Boolean algebra Bn (T) where the Boolean operations are given by [φ]·[ψ] = [φ∧ψ] [φ] + [ψ] = [φ ∨ ψ] −[φ] = [¬φ] 1=[v1 =v1],0=[v1/=v1].

For n = 0 this Boolean algebra is sometimes known as the Lindenbaum algebra of T. A complete n-type over T is essentially the same thing as an ultrafilter on Bn(T). Thus Ωn(T) is just the Stone space of Bn(T), c.f. chapter 7. Example. Suppose T admits elimination of quantifiers. Then an n-type over T is determined by the quantifier-free formulas in it (at least for $n \geq 1$). For example, let T be the theory of dense linear order without end points. A complete n-type over T is determined by specifying the order relations among v1,...,vn.

Theorem 9.13 Let A be an infinite model of a complete theory T and assume $|T| \leq \|A\|$.

(i) Given p ∈ Ωn(T) we can find an elementary extension of A of the same power as A in which p is realized.
(ii) A has an elementary extension of power max(‖A‖,|Ωn(T)|) in which every p ∈ Ωn (T) is realized.

Proof (i) Let c1, . . . , cn, a (a ∈ |A|) be new constant symbols. Since each finite subset of p is realized n A, we have that (elementary diagram of A) ∪ {φ(c1, . . . , cn) : φ ∈ p} is finitely consistent. Hence by the compactness and Löwenheim-Skolem theorems, this has a model of cardinality max(‖A‖,|p|) = ‖A‖. This is an elementary extension of A and the interpretation of c1, . . . , c_n realizes p.

Proposition 9.7 Let A be an infinite model of a complete theory T and assume $|T| \leq \|A\|$.

(i) Given p ∈ Ωn(T) we can find an elementary extension of A of the same power as A in which p is realized.
(ii) A has an elementary extension of power max(‖A‖,|Ωn(T)|) in which every p ∈ Ωn (T) is realized.

Proof (Simpson) (i) Let c1, . . . , cn, a (a ∈ |A|) be new constant symbols. Since each finite subset of p is realized n A, we have that (elementary diagram of A) ∪ {φ(c1, . . . , cn) : φ ∈ p}

is finitely consistent. Hence by the compactness and Löwenheim-Skolem theorems, this has a model of cardinality max(‖A‖,|p|) = ‖A‖. This is an elementary extension of A and the interpretation of c1, . . . , c_n realizes p.

(ii) We use an elementary chain argument. Let every p ∈ Ωn (T) is realized. n(T) = {pα : α < κ} where κ = | every p ∈ Ωn (T) is realized.

n(T)|. Put A0 = A, Aα+1 = an elementary extension of Aα, α<δ

$A\alpha$ for limit of the same power as $A\alpha$ in which $p\alpha$ is realized; $A\delta = \delta \leq \kappa$; finally $B = A\kappa$. We can show by inductionon $\alpha \leq \kappa$ that $\|A\alpha\| \leq \max(\|A\|, |\alpha|)$. Hence $\|B\| \leq \max(\|A\|, \kappa)$. If need be, use the upward Löwenheim-Skolem theorem to raise the cardinality of B to max $(\|A\|, \kappa)$.

9.7.6 SATURATED MODELS

Given a structure $B = (|B|, \Phi)$ and a set $X \subseteq |B|$, introduce new constant symbols $a(a \in X)$ and let $BX = (|B|, \Phi X)$ where $\Phi X = \Phi \cup \{(a, a) : a \in X\}$. Thus BX is just like B but with its signature expanded to include constant symbols denoting the elements of X.

Definition 9.16 Let κ be an infinite cardinal. We say that B is κ-saturated if for all $X \subseteq |B|$ of cardinality less than κ, every complete 1-type over X is realized in B.

(To be precise we should say: every complete 1-type over Th(BX) is realized in BX.)

Note that if B is κ-saturated, then for all $X \subseteq |B|$ of cardinality $< \kappa$ and all n-types p over X $(n \geq 1)$, p is realized in B.

Example. Let $B = (|B|, <)$ be a linear ordering, and let κ be an infinite cardinal. We say that B is κ-dense if for every pair of sets $X_0, X_1 \subseteq |B|$ of cardinality $<\kappa$, if $X_0, < X_1$ we can find $b \in |B|$ so that $X_0 < b < X_1$. For example, $(Q,<)$ is $\aleph 0$-dense: dense without endpoints. Also, $(R, <)$ is not $\aleph_1$-dense, as may be seen by taking $X_0 = \{0\}$, $X_1 = \{1/2, 1/4, 1/8, \ldots\}$.

Proposition 9.9 Let B be a dense linear ordering without end points. B is κ-dense if and only if B is κ-saturated.

Proof. Follows from the fact that the theory of dense linear orderings without end points admits elimination of quantifiers.

Note that a finite structure is κ-saturated for all infinite κ. If an infinite structure B is κ-saturated, then $\|B\| \geq \kappa$.

Definition 9.17 A structure B is saturated if it is $\|B\|$-saturated.

Theorem 9.14 (uniqueness of saturated models). Let A and B be saturated models of the same power. If A and B are elementarily equivalent, then they are isomorphic.

Proof. We use a back-and-forth iterations similar to the fragment consistency proof. A fix well-ordering on the order types for $||A||$ and $||B||$ to k, enumerating by stages to complete 1-typesrgument. Let $||A|| = ||B|| = \kappa$ and fix well orderings of $|A|$ and $|B|$ of order type κ. We shall define enumerations, using K-saturation for realizability, reversing the roles of A and B. The enumerations exhaust the universes of $|A|$ and $|B|$ respectively since the elements were chosen by means of fixed well orderings of order type κ.

Theorem 9.15 (universality of saturated models). Let B be a κ-saturated model and $A \equiv B$, $||A|| \leq \kappa$. Then there exists an elementary monomorphism of A into B.

Proof. Similar to the previous proof.

Example. For algebraically closed fields, theorems 8 and 9 have the fol- lowing significance. Let p = 0 or a

There is exactly one countable saturated algebraically closed field. Every countable algebraically closed field of characteristic p is embeddable into this one.

We begin with an existence theorem for countable saturated models.

Theorem 9.16 (Vaught). Let T be a complete countable theory. The following are equivalent.

(1) T has a countable saturated model;
(2) $|\Omega n(T)| \leq \aleph_0$ for all n.

Theorem 9.17 (Morley and Vaught). Let A be an infinite model and let κ be an infinite cardinal such that $\kappa \geq S_A$. Then A has a κ+-saturated elementary extension of power $||A||^{\kappa}$.

Here we visit the Löwenheim-Skolem theorm that was examined on topos in chapter 7.

Corollary 9.2 Let T be a complete theory of cardinality $\leq \kappa$, and suppose that T has an infinite model. Then T has a κ+-saturated model of power 2^k.

Proof. By Löwenheim-Skolem let A be a model of T of power κ. Apply the previous theorem to get a κ+-saturated elementary extension of power $\kappa^k = 2^\kappa$.

Corollary 9.3 Assume the Generalized Continuum Hypothesis. Let T be a complete theory of cardinality $\leq \kappa$. Then T has a saturated model of power $\kappa +$. This model is unique up to isomorphism.

Proof The G.C.H. says that $2^\kappa = \kappa +$. Apply the previous corollary and the uniqueness result for saturated models.

The following provides a necessary and sufficient condition for completeness comparable to Vaught's test.

Theorem 9.18 Assume $2^{\aleph 0} = \aleph_1$. Let T be a countable complete theory with no finite models. T is complete iff any two saturated models of T of power $\aleph_1$are isomorphic.

Proof Exercises.

Theorem 9.16 (Vaught).

Let T be a complete countable theory. The following are equivalent.

(1) T has a countable saturated model;

(2) $|\Omega n(T)| \leq \aleph_0$ for all n.

Theorem 9.17 (Morley and Vaught). Let A be an infinite model and let κ be an infinite cardinal such that $\kappa \geq S_A$. Then A has a κ+-saturated elementary extension of power $\|A\|^\kappa$.

Here we visit the Löwenheim-Skolem theorem that was examined on topos in Chapter 7.

Corollary 9.2 Let T be a complete theory of cardinality $\leq \kappa$, and suppose that T has an infinite model. Then T has a κ+-saturated model of power 2^k.

Proof By Löwenheim-Skolem let A be a model of T of power κ. Apply the previous theorem to get a κ+-saturated elementary extension of power $\kappa^k = 2^\kappa$.

Corollary 9.3 Assume the Generalized Continuum Hypothesis. Let T be a complete theory of cardinality ≤ κ. Then T has a saturated model of power κ +. This model is unique up to isomorphism.

Proof The G.C.H. says that $2^\kappa = \kappa +$. Apply the previous corollary and the uniqueness result for saturated models.

The following provides a necessary and sufficient condition for completeness comparable to Vaught's test.

Theorem 9.18 Assume $2^{\aleph_0} = \aleph_1$. Let T be a countable complete theory with no finite models. T is complete iff any two saturated models of T of power $\aleph_1$ are isomorphic.

Proof. Exercises.

9.7.7 GENERIC FILTERS ON HORN MODELS

Proposition 9.10 Let *I* be the set *T*∗. Let ϕ(*x*1 ...*xn*) be a Horn formula and let R*i i*ε*I* be models for language *L*, let *a*1 ...*an* ε Π*i* ε*I Ai* . The R*I* are fragment Horn models. If {*i*ε*I* : R*i* ϕ[(*a*1(*i*) ...*an* (*i*)]} then the direct product Π_D R_I

ϕ[*a*1*D*...*an D*], where *D* is the generic filter on *T*∗.

Completabililty theorem at *T*∗ on universal Horn sentences might be provable without *CH* : 2 $^{\omega w}$= w+. That is addressed on an alternate chapter and exercises.

Applying positive local realizability from the authorr's (Nournani 2007) we have:

Theorem 9.19 Let (*P*,≤) be a positive Horn Poset and *p*ε*P*. If *D* is a countable family of dense subsets of *P* then *T*∗(*P*) is *D*-generic filter *F* in *P* such that *p*ε*F* and every *p*ε*F* has a positive local realization.

The above theorem is a Horn densithy counterpart ot the Rasiowa–Sikorski lemma.

Proof Positive forcing theorems in this chapter, generic filter base density, and genericity (c.f. chapter 10 on Rasiowa–Sikorski).

9.8 UNIFORM AND COUNTABLY INCOMPLETE ULTRAFILTERS

From now on, we will focus on ultrafilters which are uniform and countably incomplete. In this section we explain why.

Definition 9.19 An ultrafilter U over I is uniform if every $X \in U$ has cardinality

$|X| = |I|$.

If I is a singleton $I = \{i0\}$, then $\{I\}$ is a uniform ultrafilter over I. But if I is a finite set of cardinality $|I| > 1$, then every ultrafilter over I is principal, so there is no uniform ultrafilter over I.

If I is infinite, then the set $F = \{X \subseteq I : |I \setminus X| < |I|\}$ of subsets with small complements is a proper filter over I, and an ultrafilter U over I is uniform if and only if U contains F.

From the theorems F can be extended to an ultrafilter over I, so there exist uniform ultrafilters over I.

For ultraproducts, we can always replace a non-uniform ultrafilter by a uniform ultrafilter. Suppose U is a non-uniform ultrafilter over I, and let J be an element of U of minimum cardinality. Then the set $V = U \cap P(J)$ is a uniform ultrafilter over J, and every ultraproduct $\Pi_U Ai$ is isomorphic to the ultraproduct $\Pi_V Aj$ by the mapping $g P_U V \rightarrow (gP_U J)V$.

Definition 9.20 An ultrafilter U is countably complete if U is closed under countable intersections. U is countably incomplete if U has a countable subset V suchthat $V = \emptyset$.

As with sets, the ultrapower of an L-structure A modulo U is defined as the ultraproduct $\Pi_U A = \Pi_U Ai$ where $Ai = A$ for each $i \in I$. The natural embedding is the mapping $d : A \rightarrow U A$ such that d(a) is the U-equivalence class of the constant function with value a. Note that an ultrafilter U is countably incomplete if and only if it is not countably complete. Every principal ultrafilter is countably complete. However, the hypothesis that there exists a non-principal countably complete ultrafilter is a very strong axiom of infinity that is not provable from ZFC.

The first cardinal κ such that there is a non- principal countably complete ultrafilter over a set of cardinality κ is called the first measurable cardinal. Countably complete ultraproductssatisfy ananalogues to Loś' Theorem for the infinitary logic with conjunctions and quantifiers of length $< \kappa$. It follows that when U is a countably complete ultrafilter and the cardinality of A is less than the first measurable cardinal, the ultrapower U A is trivial, that is, d : A ~= U A.

The study of countably complete ultrapowers belongs to the theory of large cardinals.

In this section we state some specifics that hold for all countably incomplete ultrafilters. The following easy result shows that countably incomplete ultrapowers of infinite structures are always non-trivial.

Proposition 9.10 Let U be a countably incomplete ultrafilter over I and let A be infinite. Then d maps A properly into the ultrapower UA, and hence ΠU A is isomorphic to a proper elementary extension of A.

Here are basic on the cardinalities of ultraproducts.

Theorem 9.20 (Frayne, Morel and Scott) Let U be a countably incomplete ultrafilter. Then U Ai is either finite or of cardinality $\geq 2^{\omega}$. Thus an ultraproduct U Ai is never countably infinite.

9.8.1 MORE ON ULTRAPRODUCTS AND ULTRAFILERS

Here is a property of countably incomplete ultraproducts which is used in many applications, such as the Loeb measure in probability theory, and the nonstandard hull of a Banach space.

Theorem 9.21 (Keisler). Suppose L is countable and U is a countably incom- plete ultrafilter over I. Then every ultraproduct UA_i is ω_1-saturated.

One advantage of ultrapowers is that they always produce complete embeddings in the following sense.

Definition 9.21 We say that mapping $h : A \rightarrow B$ is a complete embedding of A into B if for every expansion of A' of A there is an expansion B' of B such that $h : A' \prec B'$. B is a complete extension of A if $A \subseteq B$ and the identity mapping $\iota : A \rightarrow B$ is a complete embedding.

Note that every complete embedding is an elementary embedding. By Proposi- tions above, the natural embedding $d : A \rightarrow U A$ is a complete embedding, and hence every ultrapower embedding is a complete embedding.

The converse of this fact is false—there are complete embeddings which are not ultrapower embeddings (see [9], Exercise 9.4.6). However, the next result shows that each complete embedding is locally an ultrapower embedding.

Theorem 9.22 Suppose $h : A \rightarrow B$ is a complete embedding. Then for each finite subset S of B there is a $C \prec B$ such that $S \subseteq C$ and $h : A \rightarrow C$ is an ultrapower embedding.

This is a consequence of a stronger result, which states that $h : A \rightarrow B$ is a complete embedding if and only if it is a limit ultrapower embedding-limit ultrapowers are generalizations of ultrapowers which share many of their propertie).

The following two results do not mention ultrapowers but are proved using ul- trapowers.

The next result improves the classical upward Löwenheim-Skolem-Tarski theorem when the vocabulary L is large.

Theorem 9.23 Suppose κ is infinite and less than the first measurable cardinal. Then every structure of cardinality κ (with any number of relations) has an elementary extension of cardinality λ if and only if $\lambda \geq \kappa\omega$.

It is natural to ask: When is UA κ+-saturated? That is addressed on the following that the answer depends only on the complete theory of A.

Theorem 9.24 (Keisler) Let U be a regular ultrafilter over a set of cardinality κ. If $|L| \leq \kappa$ and $A \equiv B$, then UA is κ+-saturated if and only if U B is κ+ -saturated.

Let us say that a regular ultrafilter U over a set of cardinality κ saturates a complete theory T if for every model A of T, U A is κ+-saturated.

By preceding theorems, it does not matter which model of T we take. Given two complete theories S, T with countable vocabularies, we write S $\triangleright$ T if every regular ultrafilter which saturates T saturates S. This relation can be used to classify complete theories. Intuitively, higher theories in this ordering are more complex than lower ones. The $\triangleright$-class of T is the set of all S such that S $\triangleright$ T and T $\triangleright$ S. It is clear that $\triangleright$ is is reflexive and transitive, so it induces a partial order on the $\triangleright$-classes.

9.9 FUNCTORIAL PROJECTIVE SET MODELS AND SATURATION

The author had defined specific realizabilities on positive formulas since 2005. Let $L_{P,\omega}$ be the positive fragment obtained from the Keisler fragment. Define the category $L_{P,\omega}$ to be the category with objects positive fragments and arrows the subfoumual preorder on formulas. From author ASL _Sofia:

Let us define a discrete topology on the Kiesler fragment K, on $L_{1.}$. Let M be the infinite product copies on K. Give K the product toplogy. Let F = $\{K, K^2, \ldots\}$.

Observations: (i)A subset of elements of F from a topological space with a pointset topology. (ii) M is homomorphic to its product with itself. Definition If A M^{n+1} the fragment projection of A is { <1,…,n, > A)}

From authors,s Summer colloquium Sophia:

Theorem 9.25 On Fragment consistency theorem, consider the interim fragment models Mi. M |= iff Mi |= a projection on fragments sets that positively locally realize is the set modeled at the ith tower iteration.

The new areas (Nourani 2012) are as follows:

Theorem 9.27 There is an ultrafiler $I_{F\ \text{such}}$ that the fragment model ultra-prodcut UAi is a model of $L_{P,\omega}$. iff Ai |= a projection on fragments sets that Horn locally realize Ai.

Proof The latter condition guarantees that the Ai's that the ultrafilter condition is met on UAi at Los theorem:

Definition 9.22 Let us say that model M is positively saturated if for every subset X of of cardinality < k , every type is positively locally realized in M .

Corollary 9.3 The interim models Mi, tare positively k-saturated at the interim language fragment.

Corollary 9.4 Each fragment model has

(i) a k+ saturated elementary extension that is the ultraproduct on fragments k-saturated at the positive interim language.
(ii) The language has a model iff the iff every finite fragment group elementary extensions have a reduced product filter to the fragment model projections.

9.9.1 FILTERS, MODELS, AND TOPOLOGY

Here we further glimpse on topological structures from the fiter and projections view point. For any filter *F* on a set *S*, the set function defined by is finitely additive — a "measure" if that term is construed rather loosely. Therefore the statement can be considered somewhat analogous to the statement that φ holds "almost everywhere". That interpretation of membership in a filter is used (for motivation, although it is not needed for actual *proofs*) in the theory of ultraproducts in model theory, a branch of mathematical logic.

In topology and analysis, filters are used to define convergence in a manner similar to the role of sequences in a metric space. In topology and related areas of mathematics, a filter is a generalization of a net. Both nets and filters provide very general contexts to unify the various notions of limit to arbitrary topological spaces. A sequence is usually indexed by the natural numbers, which are a totally ordered set. Thus, limits in first-countable spaces can be described by sequences. However, if the space is not first-countable, nets or filters must be used. Nets generalize the notion of a sequence by requiring the index set simply be a directed set. Filters can be thought of as sets built from multiple nets. Therefore, both the limit of a filter and the limit of a net is conceptually the same as the limit of a sequence.

Let X be a topological space and x a point of X. Take N_x to be the **neighbourhood filter** at point x for X. This means that N_x is the set of all topological neighbourhoods of the point x. It can be verified that N_x is a filter. A **neighbourhood system** is another name for a **neighbourhood filter**. To say that N is a **neighbourhood base** at x for X means that for all $V_0 \in N_x$, there exists a $N_0 \in N$ such that $N_0 \subseteq V_0$. Note that every neighbourhood base is a filter base. Let X be a topological space and x a point of X. To say that filter base B **converges** to x, denoted $B \rightarrow x$, means that for every neighbourhood U of x, there is a $B_0 \in B$ such that $B_0 \subseteq U$. In this case, x is called a limit of B and B is called a **convergent filter base**. For every neighbourhood base N of x, $N \rightarrow x$.

If N is a neighbourhood base at x and C is a filter base on X, then $C \rightarrow x$ if and only if C is finer than N. For $Y \subseteq X$, to say that p is a limit point of Y in X means that for each neighborhood U of p in X, $U \cap (Y - \{p\}) \neq \emptyset$.

For $Y \subseteq X$, p is a limit point of Y in X if and only if there exists a filter base B on $Y - \{p\}$ such that $B \rightarrow p$.

For $Y \subseteq X$, the following are equivalent:

(i) There exists a filter base F whose elements are all contained in Y such that $F \rightarrow x$.

(ii) There exists a filter F such that Y is an element of F and $F \rightarrow x$.

(iii) The point x lies in the closure of Y.

Indeed:

(i) implies (ii): if F is a filter base satisfying the properties of (i), then the filter associated to F satisfies the properties of (ii).

(ii) implies (iii): if U is any open neighborhood of x then by the definition of convergence U is an element of F; since also Y is an element of F, U and Y have nonempty intersection.

(iii) implies (i): Define . Then F is a filter base satisfying the properties of (i).

9.9.2 TOPOLOGICAL STRUCTURES

Every subset of a topological space can be given the subspace topology in which the open sets are the intersections of the open sets of the larger space with the subset. For any indexed family of topological spaces, the product can be given the product topology, which is generated by the inverse images of open sets of the factors under the projection mappings. For example, in finite products, a basis for the product topology consists of all products of open sets. For infinite products, there is the additional requirement that in a basic open set, all but finitely many of its projections are the entire space.

A quotient space is defined as follows: if X is a topological space and Y is a set, and if $f: X \to Y$ is a surjective function, then the quotient topology on Y is the collection of subsets of Y that have open inverse images under f. In other words, the quotient topology is the finest topology on Y for which f is continuous. A common example of a quotient topology is when an equivalence relation is defined on the topological space X. The map f is then the natural projection onto the set of equivalence classes.

The Vietoris topology on the set of all nonempty subsets of a topological space X, named for Leopold Vietoris, is generated by the following basis: for every n-tuple $U_1, \ldots, U_n$ of open sets in X, we construct a basis set consisting of all subsets of the union of the U_i which have nonempty intersection with each U_i.

Topological spaces can be broadly classified, up to homeomorphism, by their topological properties. A topological property is a property of spaces that is, invariant under homeomorphisms. To prove that two spaces are not homeomorphic it is sufficient to find a topological property, which is not shared by them. Examples of such properties include connectedness, compactness, and various separation axioms.

For any algebraic objects we can introduce the discrete topology, under which the algebraic operations are continuous functions. For any such structure which is not finite, we often have a natural topology which is compatible with the algebraic operations in the sense that the algebraic operations are still continuous. This leads to concepts such as topological groups, topological vector spaces, topological rings and local fields.

Specialization preorder. In a space the specialization (or canonical) preorder is defined by $x \leq y$ if and only if $\mathrm{cl}\{x\} \subseteq \mathrm{cl}\{y\}$. In topology a **product space** is the cartesian product of a family of topological spaces equipped with a natural topology called the **product topology**. This topology differs from another, perhaps more obvious, topology called the box topology, which can also be given to a product space and which agrees with the product topology when the product is over only finitely many spaces. However, the product topology is "correct" in that it makes the product space a categorical product of its factors, whereas the box topology is too fine; this is the sense in which the product topology is "natural".

Definition 9.23 Given X such that or the (possibly infinite) Cartesian product of the topological spaces X_i, indexed by , and the canonical projections $p_i : X \to X_i$, the product topology on X is defined to be the coarsest topology (i.e. the topology with the fewest open sets) for which all the projections p_i are continuous. The product topology is sometimes called the Tychonoff topology.

The open sets in the product topology are unions (finite or infinite) of sets of the form $\Pi\ U_i$, where $U_i \neq X_i$ only finitely many times. The product topology on X is the topology generated by sets of the form $p_i^{-1}(U)$, where i in I and U is an open subset of X_i. In other words, the sets $\{p_i^{-1}(U)\}$ form a subbase for the topology on X. A subset of X is open if and only if it is a (possibly infinite) union of intersections of finitely many sets of the form $p_i^{-1}(U)$. The $p_i^{-1}(U)$ are sometimes called open cylinders, and their intersections are cylinder sets.

Examples ara as follows: Start with the standard topology on the real line **R** and defines a topology on the product of n copies of **R** in this fashion, one obtains the ordinary Euclidean topology on $\mathbf{R}^n$. The Cantor set is homeomorphic to the product of countably many copies of the discrete space $\{0,1\}$ and the space of irrational numbers is homeomorphic to the product of countably many copies of the natural numbers, where again each copy carries the discrete topology.

On intial topology the example is a product space product space X, together with the canonical projections. The product space is a product in the category of topological spaces. If follows from the above universal property that a map $f: Y \to X$ is continuous if and only if $f_i = p_i \circ f$ is continuous for all i in I. In many cases it is often easier to check that the component functions f_i are continuous. Checking whether a map $g : X \to Z$ is continuous is usually more difficult; one tries to use the fact that the p_i are continuous in some way.

In addition to being continuous, the canonical projections $p_i: X \to X_i$ are open maps. This means that any open subset of the product space remains open when projected down to the X_i. The converse is not true An important theorem about the product topology is Tychonoff's theorem: any product of compact spaces is compact. This is easy to show for finite products, while the general statement is equivalent to the axiom of choice.

The axiom of choice is equivalent to the statement that the product of a nonempty collection of nonempty sets is nonempty. The proof is easy enough: one needs only to pick an element from each set to find a representative in the product. Conversely, a representative of the product is a set, which contains exactly one element from each component. The axiom of choice occurs more generally in product spaces; for example, Tychonoff's theorem on compact sets is a more complex and subtle example of a statement that is, equivalent to the axiom of choice.

9.9.3 SATURATION AND MODELS

In this section we present a typical application of saturated models, namely to the proofs of "preservation theorems." Preservation theorems relate model theoretic properties of T to syntactic properties of axioms for T.

The following syntactical classification of formulas is occasionally useful.

Definition 9.24 A formula $\psi(y)$ is Σ k if it is of the form of k alternating blocks of quantifiers followed by a quantifier free matrix, and the first block is existential. The class of Πk formulas is defined similarly except that the first block is universal.

Let Γ be a class of formulas (e.g. $\Gamma = \Sigma_k$ or Π_k). A Γ-theory is a theory T such that $T \cap \Gamma$ is a set of axioms for T. We write $A\,\Gamma_{\text{ç}}\,B$ to mean that

Signature (A) = signature (B) and $Th(A) \cap \Gamma \subseteq Th(B) \cap \Gamma$.

Lemma (localization lemma). Let Γ be a class of formulas which is closed under disjunction. Let T be a theory such that the class of models of T is closed under i.e. $A \in$ Models for (T), $A\,\Gamma_{\text{ç}}\,B \in$ Models for (T). Then T is a Γ-theory.

i.e. $A \in$ Models for (T), $A\,\Gamma_{\text{ç}}\,B \in$ Models for (T). Then T is a Γ-theory.

Let us consider Tarksi's preservation theorem.

Theorem (Tarski). Suppose that T is preserved under substructures, i.e. every substructure of a model of T is a model of T. Then T is a universal, i.e., a Π_1-theory.

Theorem (Tarski) Suppose that T is preserved under substructures, that is, every substructure of a model of T is a model of T. Then T is a universal, that is, a Π_{1-}theory.

The notion of saturated model is dual to the notion of prime model in the following way: let *T* be a countable theory in a first-order language (that is, a set of mutually consistent sentences in that language) and let *P* be a prime model of *T*. Then *P* admits an elementary embedding into any other model of *T*. The equivalent notion for saturated models is that any "reasonably small" model of *T* is elementarily embedded in a saturated model, where "reasonably small" means cardinality no larger than that of the model in which it is to be embedded. Any saturated model is also homogeneous. However, while for countable theories there is a unique prime model, saturated models are necessarily specific to a particular cardinality. Given certain set-theoretic assumptions, saturated models (albeit of very large cardinality) exist for arbitrary theories. For λ, stable theories, saturated models of cardinality λ exist.

9.10 ULTRAPRODUCTS AND ULTRAFLITERS

9.10.1 POSITIVE CATEGORIES AND CONSISTENCY MODELS

Let $L_{P,\omega}$ be the positive fragment obtained from Kiesler fragment.

Define the category $L_{P,\omega}$ to be the category with objects positive fragments and arrows the subfoumual preorder on formulas.

Define a functor F: $L_{P,\omega}^{Op} \rightarrow$ Set by a list of sets Mn and functions fn. The functor F is a list of sets Fn, consisting of:

(a) the sets corresponding to an initial structure on $L_{P,\omega}$, for example the free syntax tree structure, where to f(t1, t2,.. tn) in $L_{P,\omega}$ there corresponds the equality relation f(t1, …, tn)=ft1…tn in Set;

(b) the functions fi:Fi+1 → Fi.

The following fragment theorem and proofs are written out explicit to indicate where ultrafilter realizability on ultraproducts are being applied to stratified infinite completeness and consistency on functorial models.

Theorem 9.10 Infinitary Fragment Consistency on Algebras

Let T<i, i+1> be complete theories for L<i, i+1> = Li intersect Li+1. Let Ti and Ti+1 be arbitrary consistent positive theories for the subfragments Li and Li+1, respectively, satisfying Ti contains T<i, i+1> and Ti+1 contains T<i, i+1>. Let Ai be A(Fi) and Bi be A(Fi+1).

(i) There are iterated elementary extensions Bi+1> Bi and an embedding fi: Ai < Bi

(ii) Slalom between the language pairs Li Li+1 gates to a limit model for L, a model M for $L_{P,\,\omega}$.

Proof Starting with a basis model $\mathfrak{R}_{<0,1>\text{ and}}$ $B_{<0,1>,}$ models of T_0 and T_1. $\mathfrak{R}_0$|L<0,1> and B_0|L<0,1> are models of a complete theory, therefore $\mathfrak{R}_0$|L<0,1> ° B_0|L<0,1>. It follows that the elementary diagram of A_0|l<0,1> is consistent with the elementary diagram of B_0|L<0,1>. Let T<i, i+1> be the positive theory in L<i, i+1> = Li intersect Li+1 defined by T* (A(Fi)) intersect A(Fi+1)), where T* is the theory obtained from the positive theory T augment with inductive consequences. We shall construct a limit model M realizing (i) enroute. Every Σ(x1…xn) a set of formulas of L<0,1> that is, provable in T, T*(A_0|L<0,1>) positively locally realizes Σ, that is, there is a formula φ (x1…xn) in L<01> s.t. φ is consistent with T*(A_0|L<0,1>) and for all σ ε £, T* ^ φ or T* {σ} is not consistent. Every finite subset Σ of T*(A_0|L<0,1>) has a model, therefore T*(A0|L<0,1>) has a model M_0, where for every φ ε L <0,1> T*(A0|L<0,1>) ├ φ iff M0 ╞ φ. Starting with a basis models $\mathfrak{R}_{0\text{ and}}$ B_0, we construct a positive tower. On the iteration we realize that by s preceding proposition at each stage there are elementary extensions to $\mathfrak{R}_{i\text{ and}}$ B_i there are elementary extensions 𝔑 < 𝔑,' |B < |B' such that |B' is a homomorphic image of 𝔑'. Therefore, there are elementary extensions B1> B0 and an embedding f1: A0 < B1 at L<0,1>. Passing to the expanded language L<0,1>A0, we have (A0, a)aε A0 ≡ L<0,1> A0 (B1, fa) a ε A0. g1 inverse is an extension of f1.

Iterating, we obtain the tower depicted sideways.

```
AO   <     A1   <     A2  < .........
    \  f1    I g1   \ f2    I g2
BO  <      B1   <   b2   < .....
```

Slalom between the language pairs Li Li+1 gates to a limit model for L. For each m, fm ⊂ inverse (gm) ⊂ fm+1, fm: Am–1 < Bm at L<m–1, m>. Let A = ∪ Am, m < w, B= ∪ Bm, m <w. B is isomorphic to a model B' such that A|L = B'|L. Piecing A and B' together we obtain a model M for $L_{P, w}$.

Define the category $L_{H, \omega}$ to be the category with objects Horn fragments and arrows the subfoumual preorder on formulas.

Op

Define a functor F: $L_{H, \omega} \rightarrow$ Set by a list of sets Mn and functions fn. The functor F is a list of sets Fn, consisting of:

(a) the sets corresponding to an initial structure on $L_{P, \omega}$, for example the free syntax tree structure, where to $f(t_1, t_2, .. t_n)$ in $L_{P, \omega}$ there corresponds the equality relation $f(t_1, \ldots, t_n)=ft_1 \ldots t_n$ in Set;

(b) the functions fi:Fi+1 → Fi.

Proposition 9.6 Infinitary Horn Fragment Consistency

Let T<i, i+1> be the complete theory in L<i, i+1> = Li intersect Li+1, defined by Th (A(Fi) intersect A(Fi+1)), Let Ti and Ti+1 be arbitrary consistent positive Horn theories for the subfragments Li and Li+1, respectively, satisfying Ti contains T<i, i+1> and Ti+1 contains T<i, i+1>. Let Ai be A(Fi) and Bi be A(Fi+1). Starting with a basis model A0, A0|L<0,1> and B0|L<0,1> are models of a complete theory, hence, A0|L<0,1> ≡ B0|l<0,1>. It follows that the elementary diagram of A0|l<0,1> is consistent with the elementary diagram of B0|L<0,1>.

(i) There are iterated elementary extensions Bi+1> Bi and an embedding fi: Ai < Bi

(ii) Slalom between the language pairs Li, Li+1 gates to a limit model for L, a model M for L_H^{ω}.

Define the category $L_{H, \omega}$ to be the category with objects Horn fragments and arrows the subfoumual preorder on formulas.

Define a functor F: $L_{H, \omega}^{Op} \rightarrow$ Set by a list of sets Mn and functions fn. The functor F is a list of sets Fn, consisting of:

(a) the sets corresponding to an initial structure on $L_{P, \omega}$, for example the free syntax tree structure, where to $f(t_1, t_2, ..$ tn) in $L_{P, \omega}$ there corresponds the equality relation $f(t_1, \ldots,$ tn)=$ft_1 \ldots$tn in Set;

(*b*) the functions fi:Fi+1 → Fi. ⊆

9.10.2 PROJECTIVES IN CATEGORIES

In category theory, the notion of a **projective object** generalizes the notion of a projective module. An object *P* in a category **C** is **projective** if the hom functor Hom (P, —) : C → Set preserves epimorphisms. That is, every morphism *f:P→X* factors through every epi: *Y→X.*

Let C be an abelian category. Here an object P ∈ C is called a *projective object* if Hom (P, —) : C → Ab is an exact functor, where Ab is the category of abelian groups. The dual notion of a projective object is that of an **injective object**: An object W in an abelian category C is *injective* if the Hom (—,W) functor from C to Ab is exact. Let A be an abelian category. A is said to have *enough projectives* if, for every object A of A, there is a projective object P of A and an exact sequence P →A → 0. That is the map P →A is an epimorphism.

Let R be a ring with 1. Consider the category of left R-modules M_R . M_R is an abelian category. The projective objects in M_R are precisely the projective left R-modules. So R is itself a projective object in M_R . Dually, the injective objects in M_R are exactly the injective left R-modules.
The category of left (right) R-modules also has enough projectives. This is true since, for every left (right) R-module M, we can take F to be the free (and hence projective) R-module generated by a generating set X for M (we can in fact take X to be M). Then the canonical projection p : F → M is the required surjection.

Exmples What are the projective objects in the category of rings with identity {0} and Z . If R is projective and Z⟨R⟩ is the free ring generated by the *set* R then there is a surjective ring hom. κ:Z⟨R⟩→R and, by projectivity, a ring hom. i:R→Z⟨R⟩ with κ∘i=id R . In particular R has characteristic 0 and is free of zero-divisors.

9.10.3 MORE ON HORN MODELS

Barr based on Gray (1989) explores the connection between categories of models of Horn theories and models of finite limit theories that might prove an interesting application for fragment consistent Horn product models for us hindsight.

By an equational Horn theory is meant an equational theory augmented by a set of conditions of the form

$$[\varphi 1(x) = \psi 1(x)] \wedge \cdots \wedge [\varphi n(x) = \psi n(x)] \Rightarrow [\varphi(x) = \psi(x)]$$

where φ, ψ, φi and ψi are operations in the theory and x stands for an element of a product of sorts to which they all apply. Note that since we are using the whole theory (that is, the full clone), irrelevant arguments may be added to operations so that the operations in Gray (1989).

The above clause has the same arguments. Of course, if a sort is empty, any equational sentence stated in terms of that sort is satisfied.

A generalized equational Horn theor is an equational theory augmented by a set of conditions of the form

$[(\varphi i(x) = \psi i(x)] \Rightarrow [\varphi(x) = \psi(x)]$, allowing a possibly infinite conjunction.

As observed, the only projectives in the category of small categories are the discrete ones, which represent the set of objects

Definition 9.25 A formula is said to be positive iff it is built from atomic formulas using only the connectives &, v and the quantifiers∀, ∃.

Definition 9.26 A formula φ (x_1, x_2, …, xn) is preserved under homomorphisms iff for any homomorphisms f of a model A onto a model B and all a_1, …, an in A if A ⊨ [a_1, …, an] B ⊨ [fa_1, …, fan].

Theorem 9.28 A consistent theory is preserved under homomorphisms iff T has a set of positive axioms.

Positive forcing had defined T* to be T augmented with induction schemas on the Generic diagram functions. That can effectively generates Tarskian models since Tarskian presentations can be created with Skolemization on arbitrary sentences on generic diagrams with the Skolem functions instantiating the generic diagram functions.

Here, certain specifics from chapter 4 to be self-contained.

Proposition 9.6 Let R and D be models for L. Then R is isomorphically embedded in D iff D can be expanded to a model of the diagram of R.

Proposition 9.7 Let R and D be models for L. Then R is homomorphically embedded in D iff D can be expanded to a model of the positive diagram of R.

Let Σ be a set of formulas in the variables x_1…xn. Let R be a model for L. We say that R realizes Σ iff some n-tuple of elements of A satisfies Σ in R. R omits Σ iff R does not realize Σ.

For our purposes we define a new realizability basis.

Definition 9.27 Let Σ (x1...xn) be a set of formulas of L. Say that a positive theory T in L positively locally realize Σ iff there is a formula φ (x1...xn) in L s.t. (i) φ is consitent with T (ii)forallσεΣ, T ⊨ φorT∪ σisnotconsistent.

Defintion 9.28 Given models A and B, with generic diagrams DA and DB we say that DA homomrphially extends DB iff there is a homomorhic emdedding f: A B.

Consdier a complete theory T in L. A formula φ (x1,.,xn) is said to be complete (in T) iff for every formula ψ(x1,.,xn), exactly one of T ⊨ φ ψ or T ⊨ φ¬ ψ . A formula θ (x1,...,xn) is said to be completable (in T) iff there is a complete formula φ (x1,.,xn) with T models φ θ. If that can't be done θ is said to be incompletable.

Theorem 9.29 Let L1, L2 be two positive languages. Let L = L1 ∩ L2. Suppose T is a complete theory in L and T1 ⊃ T, T2 ⊃ T are consistent in L1, L2, respectively. Suppose there is model M definable from a positive diagram in the language L1 ∪ L2 such that there are models M1 and M2 for T1 and T2 where M can be homorphically embedded in M1 and M2.

(i) T1 ∪ T2 is consistent. (ii) There is model N for T1 ∪ T2 definable from a positive diagram that homomorphically extends that of M1 and M2.

Theorem 9.30 (author ASL March 07) Let L1, L2 be two positive languages. Let L = L1 ∩ L2. Suppose T is a complete theory in L and T1 ⊃ T, T2 ⊃ T are consistent in L1, L2, respectively. Then

(i) T1 ∪ T2 has a model M, that is a positive end extension on Models M1 and M2 for T1, and T2, respectively si
(ii) M is definable from a positive diagram in the language L1 ∪ L2.

Theorem 9.31 Considering a Tarskian presentation P for a theory T that has a positive local realization, with T*.

Proof Follows from positive local realizability on the generic omitting type T*.

9.10.4 POSITIVE CATEGORIES AND CONSISTENCY MODELS

During 2005–2007 the author explored positive and Horn fragment categories, based on his dissertation that proved to be on intuitionist forcing, on positive forcing on Kiesler fragments. Positive forcing had defined T * on a theory T to be T augmented with induction schemas on the generic diagram functions, for a model M fo T. Let P be a poset, F a family of sets, G subset of P. The author defined positive local realizability on 2005–2007 publications (ASL).

Theorem 9.32 The positive forcing $T*$ is a F -generic filter.

$T*$ is a principal proper filter.

Proposition 9.8 Let I be the set $T*$. Let $\phi(x1 \ldots xn)$ be a Horn formula and let Ri $i\varepsilon I$ be models for language L, let $a1 \ldots an$ ε Πi εI Ai . The RI are fragment Horn models.

If $\{i\varepsilon I$: Ri, $\phi[(a1(i) \ldots an\ (i)]\}$ then the direct product ΠD RI $\phi[a1D \ldots an\ D]$, where D is the generic filter on $T*$.

Completabililty theorem at $T*$ on universal Horn sentences might be provable without CH .

Applying positive local realizability from 2, 3:

Theorem 9.33 Let $(P,\leq)$ be a positive Horn Poset and $p\varepsilon P$. If D is a countable family of dense subsets of P then $T*(P)$ is D-generic filter F in P such that $p\varepsilon F$ and every $p\ \hat{I}\ F$ has a positive local realization.

The above theorem is a Horn density counterpart to the Rasiowa–Sikorski lemma. That will be examined further in chapter 10.

The above theorem is a Horn density counterpart to the Rasiowa–Sikorski lemma. That will be examined further in chapter 10.

9.11 EXERCISES

1. Let I be the set $T*$. Let $\phi(x1 \ldots xn)$ be a Horn formula and let Ri $i \in I$ be models for language L, let $a1 \ldots an \in \Pi i \in I\, Ai$. The R$I$ are fragment Horn models.
   If $\{i \varepsilon I$: Ri
   $\phi[(a1(i) \ldots an\,(i)]\}$ then the direct product Π_D RI
   $\phi[a1D \ldots an\, D]$, where
   D is the generic filter on $T*$.
   Completabililty theorem at T^* on universal Horn sentences might be provable without CH: 2 = +.
2. T* is a principal proper filter.
3. Let M ∈ S. Then M is contained in a K-generic structure M* ∈ S. *Hint*: Use Lindenbaum's lemma.
4. Let M and M' be K-generic structures in $S_{K \text{ such that}}$ M' ⊃. Then M's is an elementary extension of M.
5. Consider a model M, and the type $\{x \neq m : m \in M\}$.
   (i) Each finite subset of this type is realized in the (infinite) model M, show that is trivially not realized.
   (ii) Show that parameter sets must be which are strictly smaller than the model to have saturation on infinite models.
6. Prove the second half on theorem 9.n? using e an elementary chain argument. Let $\Omega n(T) = \{p\alpha : \alpha < \kappa\}$ where $\kappa = |\Omega n\,(T)|$. Put $A0 = A$, $A\alpha+1$ = an elementary extension of $A\alpha$.
7. What are the projective objects in the category of rings with identity.
8. (i) A structure is consistent with K iff it is consistent with $K_\forall$.
   (ii) For $K_\forall$ every substructure of a models of K satisfies all sentences of $K_\forall$ which are defined in it.

9. Let M be K-generic structure and let M' $\supset$ M be a model of K. Let X be any $\forall\exists$ sentence defined in M'. The x holds in M. Hint:Apply the preservation theorem.
10. Prove that a theory is preserved under substructures iff is is a universal theory.
    Hint: Following theorem 9.13's outline with compactness.
11. Show that the countable random graph, with the only non-logical symbols being the edge existence relation, is saturated.
    Hint: Consider the finite subgraph consisting of the variables and parameters used to define the type.
12. Assume the Generalized Continuum Hypothesis. Let T be a complete theory of cardinality $\leq \kappa$. Then T has a saturated model of power κ +. This model is unique up to isomorphism.
13. Assume $2^{\aleph 0} = \aleph_1$. Let T be a countable complete theory with no finite models. T is complete iff any two saturated models of T of power $\aleph_1$ are isomorphic.
14. Prove that for a theory T that is countable, complete and has an infinite model, is $\aleph_0$-categorical if and only if $\forall$n(Ωn(T) is finite).
15. Let T be a complete theory and let p be a complete n-type over T, n $\geq$ 1. Prove the following:
    (i) Every finite subset of p is realized in every model of T.
    (ii) p might not be realized in a given model of T. For example, let T = T0 = the theory of algebraically closed fields of characteristic zero.
    Hint: Use a 1-type p at rational fields.
16. Prove that every complete embedding is an elementary embedding. By Propositions above, the natural embedding d : A $\to$ U A is a complete embedding, and hence every ultrapower embedding is a complete embedding.

KEYWORDS

- **projective determinants**
- **filter base of tails**
- **sunstructures**
- **countably saturated**

CHAPTER 10

A GLIMPSE ON ALGEBRAIC SET THEORY

CONTENTS

10.1 PRELIMINARIES

The aim of algebraic set theory is to provide a uniform categorical semantics for set theories of different kinds (classical or constructive, predicative or impredicative, well-founded or nonwell-founded, etc.). We will describe what the basis algebraic set theory is from Godel to now. The base area ranges from ultraproducts and ultrafilters to topological filters based on the preceding chapters.

In the foundations of mathematics, von Neumann–Bernays–Gödel set theory (NBG) is an axiomatic set theory that is a conservative extension of the canonical axiomatic set theory ZFC. A statement in the language of ZFC is provable in NBG if and only if it is provable in ZFC. The ontology of NBG includes proper classes, objects having members but that cannot be a member of other entities. NBG's principle of class comprehension is predicative; quantified variables in the defining formula can range only over sets. Allowing impredicative comprehension turns NBG into Morse-Kelley set theory (MK). NBG, unlike ZFC and MK, can be finitely axiomatized.

The defining aspect of NBG is the distinction between proper class and set. Let *a* and *s* be two individuals. Then the atomic sentence a ∈ S is defined if *a* is a set and *s* is a class. In other words, *a* ∈ S is defined unless *a* is a proper class. A proper class is very large; NBG even admits of "the class of all sets", the universal class called V. However, NBG does not admit "the class of all classes" (which fails because proper classes are not "objects" that can be put into classes in NBG) or "the set of all sets" (whose existence cannot be justified with NBG axioms).

By NBG's axiom schema of Class Comprehension, all objects satisfying any given formula in the first order language of NBG form a class; if the class would not be a set in ZFC, it is an NBG proper class. The development of classes mirrors the development of naive set theory. The principle of abstraction is given, and thus classes can be formed out of all individuals satisfying any statement of first order logic whose atomic sentences all involve either the membership relation or predicates definable from membership. Equality, pairing, subclass, and such, are all definable and so need not be axiomatized — their definitions denote a particular abstraction of a formula. Classes lacking representations, such as the class of all sets that do not contain themselves (the class invoked by the Russell paradox), are the proper classes.

John von Neumann in 1938 stated his axioms and showed they were adequate to develop set the ∈-relation. In contrast to Fraenkel and Skolem, von Neumann found the axiom of replacement so essential to his work that he stated that no theory of ordinals is possible at all without this axiom. A criterion identifying classes that are too large to be sets. Zermelo did not provide such a criterion. His set theory avoids the large classes that lead to the paradoxes, but it leaves out many sets, such as the one mentioned by Fraenkel and Skolem. Von Neumann's criterion is: A class is too large to be a set if and only if it can be mapped onto the universal class. Von Neumann realized that the paradoxes could be avoided by not allowing such large classes to be members of any class. Combining this restriction with his criterion, he obtained his axiom of limitation of size: A class X is not a member of any class if and only if X can be mapped onto the universal class. He proved that this axiom implies the axioms of replacement and separation, and implies that the universal class can be well ordered, which is equivalent to the axiom of global choice.

Finite axiomatization. Fraenkel and Skolem formalized Zermelo's imprecise concept of "definite propositional function", that appears in his axiom of separation was developed on Skolem's axiom schema of separation that is currently used in ZFC. Von Neumann avoided axiom schemas by formalizing the concept of "definite propositional function" with his functions, whose construction requires only finitely many axioms. This led to his set theory having finitely many axioms. (Montague 1961) proves that ZFC cannot be finitely axiomatized theory. Von Neumann took functions and arguments as primitives. His functions correspond to classes and functions that can appear as arguments correspond to sets. In fact, he defined classes and sets using functions that can take only two values (that is, indicator functions whose domain is the class of all arguments).

Von Neumann's work in set theory was influenced by Cantor's articles, Both Fraenkel and Skolem pointed out that Zermelo's axioms cannot prove the existence of the set $\{Z0, Z1, Z2, \ldots\}$ where $Z0$ is the set of natural numbers, and $Zn+1$ is the power set of Zn. They then introduced the axiom of replacement, which would guarantee the existence of such sets. However, they were reluctant to adopt this axiom: Fraenkel's opinion was that Replacement was too strong an axiom for general set theory and Skolem only wrote that 'we could introduce' Replacement. Von Neumann worked on the deficiencies in Zermelo's set theory and introduced several innovations to remedy them, including a theory of ordinals. Zermelo's set theory does not contain Cantor's theory of ordinal numbers. Von Neumann recovered this theory by defining the ordinals using sets that are well-ordered by

Zermelo's set theory does not exclude nonwell-founded sets. Fraenkel and von Neumann introduced axioms to exclude these sets. Von Neumann introduced the axiom of regularity, which states that all sets are well founded. Between 1937 and 1954 Bernays using sets, following the tradition established, by Cantor, Dedekind, and Zermelo. His classes followed the tradition of Boolean algebra since modified von Neumann's theory by taking sets and classes as primitives. They permit the operation of complement as well as union and intersection. Bernays handled sets and classes in a two-sorted logic. This required the introduction of two membership primitives: one for membership in sets, and one for membership in classes. With these primitives, Bernays rewrote and simplified von Neumann's axioms. He also adopted the axiom of regularity, and replaced the axiom of limitation of size with the axioms of replacement and von Neumann's choice axiom. Von Neumann's work shows that the last two changes allow Bernays' axioms to prove the axiom of limitation of size.

Gödel (1940) simplified Bernays' theory by making every set a class, which allowed him to use just one sort for classes and one membership primitive. He also introduced a predicate indicating which classes are sets. Gödel modified some of Bernays' axioms, and introduced the axiom of global choice to replace von Neumann's choice axiom. He used his axioms in his 1940 monograph on the relative consistency of global choice and the generalized continuum hypothesis.

Gödel's approach to set theory, with its emphasis on hierarchical truth, should be set into the context of the axiomatic development of the subject. (Zermelo 1908) had provided the initial axiomatization of "the set theory of Cantor and Dedekind", with characteristic axioms Separation, Infinity, Power Set, and of course, Choice. Work most substantially of John von Neumann (1923,1928) on ordinals led to the incorporation of Cantor's transfinite numbers as now the ordinals and the axiom schema of Replace- ment for the formalization of transfinite recursion. (Von Neumann 1929) also formulated the axiom of Foundation, that every set is well-founded, and defined the *cumulative hierarchy* in his system via transfinite recursion: The axiom entails that the universe *V* of sets is globally structured through a stratificationin to cumulative "ranks" Vα, where with ℘(X)={Y |Y ⊆X} denoting the power set of *X* ,

The axiom entails that the universe *V* of sets is globally structured through a stratificationin to cumulative "ranks" Vα, where with ℘(X)={Y |Y ⊆X} denoting the power set of *X* ,

$V_0 = \varnothing$; $V_{\alpha+1} = \wp(V\alpha)$; $V = \cup_{\alpha< b} V\alpha$ for limit ordinals *b* ; and
$V = \cup \alpha V\alpha$.

Zermelo in 1930 subsequently provided his final axiomatization of set theory, proceeding in a second-order context and incorporating both Replacement, subsumming Separation, and Foundation. The standard axiomatization ZFC is the first-order version of the Zermelo [axiomatization, and ZF is ZFC without AC.

The constructible universe *L for* set theory is an independent field of mathematics by Gödel's formulation of the class *L* of *constructible* sets through which he established the relative consistency of AC and CH (Godel 1935,1937). The constructible sets are what is obtained by Russell's ramified hierarchy of types, if extended to include transfinite orders. The extension to transfinite orders has the consequence that the model satisfies the impredicative axioms of set theory, because an axiom of reducibility can be proved for sufficiently high orders."

Important to an algebraic construction of *L:*

Gödel had refined the cumulative hierarchy of sets described in his 1933 lecture to a hierarchy of *definable* sets which is analogous to the orders of Russell's *ramified* theory. Gödel assimilated the ramified theory and its motivations as of consequence and now put the theory to a new use. Gödel continued the indexing of the hierarchy through *all* the ordinals as given beforehand to get a class model of set theory and thereby to achieve *relative* consistency results. That "an axiom of reducibility can be proved for sufficiently high orders" is an allusion to how Russell's problematic axiom would be rectified in the consistency proof of CH and the axiom of Replacement.Von Neumann's ordinals would be the spun a thin hierarchy of sets, towards both the AC and CH.

In a brief 1939 account Gödel informally presented *L* much as is done today: For any set *x* let def(x) denote the collection of subsets of *x* definable over ⟨x,∈⟩ via a first-order formula allowing parameters from *x*. Then define

$L0 = \varnothing$; $L\alpha{+}1 = \mathrm{def}(L\alpha)$, $Lb = \cup\, \alpha{<}b\ L\alpha$ for limit ordinals

and the *constructible universe* $L = \cup\alpha L_{\alpha}$.

Toward the end (Gödel 1939) pointed out that *L* "can be defined and its theory developed in the formal systems of set theory themselves." This is a remarkable understatement of arguably the central feature of the construction of *L: L* is a class definable in set theory via a transfinite recursion that could be based on the formalizability of def(x), the definability of definability. Gödel had not embraced the definition of truth as itself clarificatory,16 but through his work he in effect drew it into mathematics to a new mathematical end.

However understated in Gödel's writing, his great achievement here as in arithmetic is a subtle merging of metamathematical notions into mathematics. In the proof of the incompleteness theorem, Gödel had encoded provabil-ity syntax and played on the interplay between truth and definability. Gödel now encoded satisfaction semantics with the room offered bythe transfinite indexing, making truth, now definable for levels, part of the formalism and part of the subject matter. In modern parlance, an *inner model* of ZFC is a transitive (definable) class containing all the ordinals such that, with membership and quantification restricted to it, the class satisfies eachaxiomof ZFC. Gödel ineffect argued in ZF to show that L is an inner model of ZFC, and moreover that *L* satisfies CH. He thus established the relative consistency Con(ZF) implies Con(ZFC + CH). In what follows, we describe his proofs that *L* is an inner model of ZFC and that *L* satisfies CH.

10.2 ULTRAPRODUCTS AND ULTRAFILTERS ON SETS

In applications of the ultrapower, an ultrafilter U is picked and to simultaneously takes the ultrapower of everything in sight modulo U. An efficient way to do this is to begin with a superstructure and use the ultrapower to build a nonstandard univers. Ultrapowers are also used to construct models of various nonstandard set theories. Let T be a complete theory and let κ be a cardinal. We write ν(T,k) = the number of nonisomorphic models of T of cardinality k. This function is a central object of study in pure model theory.

In this section, we study $\nu(T, \aleph_0)$ where T is complete and countable. We begin with the case when $\nu(T,\aleph_0) = 1$, i.e. T is $\aleph_0$-categorical. From the preceding chapters's notation Fn is the complete n-type. We state several well-known theorems without proof to carry on.

Theorem 10.1 (Ryll-Nardzewski). Let T be a complete countable theory with no finite models.

The following are equivalent:

(1) T is $\aleph_0$-categorical.
(2) Fn (T) is finite for each $n \in \omega$.
(3) Every countable model of T is prime.
(4) Every countable model of T is saturated.

Next we consider the situation where $1 < \nu(T,\aleph_0) < \aleph_0$. We say that a countable model A of T is weakly saturated if every $p \in \Phi n(T)$, $n \in \omega$ is realized in A.

Given a set X, the n-th cumulative power set of X is defined recursively by $V_0(X) = X$, $V_{n+1}(X) = V_n(X) \cup \wp(V_n(X))$.

The superstructure over X is the union of the cumulative power sets and is denoted by V (X),

$$V(X) = \bigcup_{0}^{\infty} V n(X).$$

The superstructure V (X) has a membership relation ∈ between elements of Vn(X) and Vn+1(X), n = 0,1,2,.... We treat the elements of X as atoms, and always assume that $\Phi \notin X$ and that no $x \in X$ contains any elements of V (X). We then consider the structure $V(X) = \langle V(X), \in \rangle$ whose vocabulary has the single binary relation ∈.

Following Kiesler's views the ultrapower construction was the motivation for the Rudin-Keisler ordering. It is a pre-ordering on the class of all ultrafilters. Heuristically, higher ultrafilters in the ordering produce larger ultrapowers with respect to elementary embeddings. Now we state a brief on ultraproducts and ultrapowers to study In applications of the ultrapower, one often picks an ultrafilter U and simultaneously takes the ultrapower of everything in sight modulo U. An efficient way to do this is to begin with a superstructure and use the ultrapower to build a nonstandard universe. We will briefly sketch how this is done, and then point out a connection between nonstandard universes and complete embeddings.

We now define the ultraproduct operation on sets. Let U be an ultrafilter over I, and for each $i \in I$ let Ai be a nonempty set. The ultraproduct Π_U Ai is obtained by first taking the cartesian product $\Pi i \in I$ Ai and then identifying two elements which are equal for U-almost all $i \in I$. The formal definition is as follows:

Definition 10.2 Let U be an ultrafilter over I. Two elements f,g of the cartesian product $i \in I$ Ai are said to be U-equivalent, in symbols $f =_U g$, if the set $\{i : f(i) = g(i)\}$ belongs to U. The U-equivalence class of f is the set $fU = \{g : f =_U g\}$. The ultraproduct Π_U Ai is defined as the set of U-equivalence classes

$$\prod_U Ai = \{f_U : f \in \prod_{i \in I} Ai, \}.$$

Given a nonempty set A, the **ultrapower** of A modulo U is the defined $i \in I$ as the ultraproduct $\Pi_U A = \Pi_U Ai$ whereAi =A for each $i \in I$. The natural embedding is the mapping $d : A \to \Pi_U A$ such that d(a) is the U-equivalence class of the constant function with value a. It is easily seen that d is injective.

Proposition 10.1 (Expansion Property) Suppose $L1 \subseteq L2$, and for each $i \in I$, Ai is an L1-structure and Bi is an expansion of Ai to L2. Then for every ultrafilter U over I, Π_U Bi is an expansion of Π_U Ai.
Proof Exercises.

Definition 10.3 We say that mapping $h : A \to B$ is a complete embedding of A into B if for every expansion of A′ of A there is an expansion B′ of B such that $h : A' < B'$. B is a complete extension of A if $A \subseteq B$ and the identity mapping $\iota : A \to B$ is a complete embedding.

Note that every complete embedding is an elementary embedding. By Proposition 10.1, the natural embedding $d : A \to P_U A$ is a complete embedding, and hence every ultrapower embedding is a complete embedding. The converse of this fact is false—there are complete embeddings which are not ultrapower embeddings. However, the next result shows that each complete embedding is locally an ultrapower embedding.

Theorem 10.2 Suppose $h : A \to B$ is a complete embedding. Then for each finite subset S of B there is a $C < B$ such that $S \subseteq C$ and $h : A \to C$ is an ultrapower embedding.

The natural Ultrafilters over N where that odereing is applied has a rich structure with promising possibilities for applications, because they give the "smallest" nontrivial ultrapowers. Given a function $f : I \to J$ and an ultrafilter U over I, we define $f[U]=\{Y \subseteq J : f-1(Y) \in U\}$. f[U] is an ultrafilter over J (Exercise 6).

Definition 10.4 Given ultrafilters U over I and V over J, $V \leq_{RK} U$ means that there exists a function $f : I \to J$ such that $V = f[U]$. We write $U \equiv_{RK} V$ if $U \leq_{RK} V$ and $V \leq_{RK} U$, and $U <_{RK} V$ if $U \leq_{RK} V$ but not $U \equiv_{RK} V$.

Note that if $V \leq_{RK} U$ then $\min\{|Y| : Y \in V\} \leq \min\{|X| : X \in U\}$. Note that $\leq_{RK}$ is transitive and symmetric. It is also upward directed, i.e., whenever V,W are ultrafilters and $U = V \times W$, we have $V \leq_{RK} U$ and $W \leq_{RK} U$. An ultrafilter U is countably incomplete if and only if there is a free ultrafilter V over N such that $V \leq_{RK} U$.

The following result was found independently by several people.

Theorem 10.3 Let U be an ultrafilter over I and let $f : I \to J$. Then we have $f[U] \equiv_{RK} U$ if and only if the restriction of f to some $X \in U$ is one to one.

Next the connection between the pre-ordering ≤RK and ultrapowers shows that higher ultrafilters with respect to $\leq_{RK}$ give bigger ultrapowers with respect to elementary embeddability.

Proposition 10.2 Let U be an ultrafilter over I and V be an ultrafilter over J. Then $V \leq_{RK} U$ if and only if for every A, $P_V A$ is elementarily embeddable in $P_U A$. Also, $V \equiv_{RK} U$ if and only if for every A, $P_V A \cong P_U A$.

A recurrent theme in the literature is to exploit the interplay between ultrafilters V ≤RK U and elements of an ultrapower modulo U.

To explain the idea, we introduce some notation and state a result. Let U be an ultrafilterover I and let $B=P_U A$. For each function $f:I \to A$, let B[f] bet the set of all elements$(g \circ f)U \in B$ where $g:A \to A$. Note that If $f : I \to A$

then $f[U] \leq_{RK} U$. If $f,h : I \rightarrow A$ and $h[U] \leq_{RK} f[U]$ then $h_U \in B[f]$. If f is a constant function, then f[U] is principal and B[f] = d(A). If the structure A has a function symbol for every $g : A \rightarrow A$, then B[f] is just the substructure of B generated by f_U.

Proposition 10.3 Suppose U is an ultrafilter over I, $\mathfrak{R} = \Pi_U A$, and f is a function from I into A.

(i) $\mathfrak{R}[f] < \mathfrak{R}$;
(ii) $\mathfrak{R}[f] \sim = P_{f[U]} A$.
(iii) If $f[U] \equiv_{RK} U$ then $\mathfrak{R}[f] = \mathfrak{R}$.
(iv) If $|I| \leq |A|$ and $\mathfrak{R}[f] = \mathfrak{R}$,t hen $f[U] \equiv_{RK} U$.

Proof (KIESLER) (i) is proved by induction on complexity of formulas. (ii) f[U] is an ultrafilter over A, and the isomorphism is given by the mapping

$(g \circ f)U \rightarrow gf[U]$. (iii) Suppose $f[U] \equiv_{RK} U$. By Theorem 8.2, there is an $X \in U$ such that the restriction of f to X is one to one. Then for any $h : I \rightarrow A$ there exists $g : A \rightarrow A$ such that $(g \circ f)U = hU$, and hence $\hat{A}[f] = \hat{A}$.

(iv) Suppose $|I| \leq |A|$ and B[f] = B. Then there is a one to one function $h: I \rightarrow A$. Since $\hat{A}[f] = \hat{A}$, $hU = (g \circ f)U$ forsome$g: A \rightarrow A$. Then $h[U] = (g \circ f)[U] = g[f[U]]$, so $h[U] \leq_{RK} f[U]$. Since h is one to one we have $U \equiv R_K h[U]$, and $f[U] \leq_{RK} U$ by definition, so $f[U] \equiv_{RK} U$.

Every principal ultrafilter is countably complete. However, the hypothesis that there exists a non-principal countably complete ultrafilter is a very strong axiom of infinity that is not provable from ZFC. The first cardinal κ such that there is a non- principal countably complete ultrafilter over a set of cardinality κ is called the first measurable cardinal. This cardinal, if it exists, is exceedingly large, for example, κ must be the κ-th inaccessible cardinal. Countably complete ultraproducts satisfy ananalogue of Lós theorem for the infinitary logic with conjunctions and quantifiers of length $< \kappa$. It follows that when U is a countably complete ultrafilter and the cardinality of A is less than the first measurable cardinal, the ultrapower $\Pi_U A$ is trivial, that is, $d : A \sim =_U A$. For this reason, the study of countably complete ultrapowers belongs to the theory of large cardinals

We conclude this section with some results which hold for all countably in- complete ultrafilters. The following easy result shows that countably incomplete ultrapowers of infinite structures are always non-trivial.

Proposition 10.4 Let U be a countably incomplete ultrafilter over I and let A be infinite. Then d maps A properly into the ultrapower $\Pi_U A$.

10.3 ULTRAFILTERS OVER N

Denoting the Stone-Cěch compactification of the discrete topology on N by β(N), defined as the topology on the set of all ultrafilters over N with as closed base the family of all sets {U ∈ β(N) : X ∈ U}, where X ∈ ℘(N). The Stone representation theorem ([52]) shows that β(N) is a compact totally disconnected Hausdorff space. One can identify each n ∈ N with the principal ultrafilter which contains the singleton {n}, so that β(N) \ N is the space of all free ultrafilters over N.

W.Rudin proves that there are free ultrafilters over N with different topological properties.

With Rudin-Kieselr ordering Ng-Render prove the folloing interesting point on a lack of countable compactness on certain U infinite topology.

Theorem 10.5 (Ng-Render) Let U be a free ultrafilter over N, let A be infinite, and let S be the topology on U A whose closed basis is the family of all sets of the form $\{f_{complete} : f \in CN\}$ where $C \subseteq A$. Then S is not countably compact.

Say that a free ultrafilter U over N is minimal if there is no free ultrafilter V over N such that $V <_{RK} U$. Note that if U is minimal then there is no free ultrafilter V over any set J such that V <RK U. Assuming Martin's axiom, Booth proved that minimal ultrafilters exist, and Blass proved that there are 2^{ω} minimal ultrafilters. Kunen proved that the existence of minimal ultrafilters is independent of ZFC. There are several natural equivalent conditions for a minimal ultrafilter.

Theorem 10.6 (Kunen, Rowbottom) Let U be a free ultrafilter over N. The following are equivalent:

(i) U is minimal;
(ii) U is selective (i.e., U contains a choice set for every partition of N into infinitely many classes which are not elements of U);
(iii) U is Ramsey (i.e., U contains a homogeneous set for every partition of [N]k into two classes, where [N]k is the set of unordered k-tuples from N)

There are also nice conditions which involve ultrapowers.

Theorem 10.7 Let A be an infinite structure, U be a free ultrafilter over N, and B= UA.

(i) U is minimal if and only if for every $f : N \rightarrow A$, either B[f] = d(A) or B[f] = B.
(ii) Suppose A has symbols for every function $g : A \rightarrow A$. Then U is minimal if and only if the only substructures of B are d(A) and B itself.

Definition10.5 A free ultrafilter U over I is called regular if there is a set $E \subseteq U$ such that $|E| = |I|$ and each $i \in I$ belongs to only finitely many $X \in E$.

Proposition 10.5 (i) If J is infinite, $I = \wp_{\omega}(J)$, and $\{i \in I : j \in i\} \in U$ for each $j \in J$, then U is regular.

(ii) There exist regular ultrafilters over each infinite set I. (iii) Every regular ultrafilter is countably incomplete and uniform. (iv) Suppose U is a regular ultrafilter over I and V is an ultrafilter over J. If $|I| = |J|$ and $U \leq_{RK} V$, then V is regular. If $|J| \leq |I|$ then $U \times V$ and $V \times U$ are regular.

It is obvious that if U is an ultrafilter over I, then any ultrapower $P_U A$ has cardinality at most |AI|. The following result shows that when U is regular, this maximum cardinality is attained.

Theorem 10.7 (Frayne, Morel and Scott) If U is a regular ultrafilter over I and A is infinite, then | U A| = |AI|.

It follows that each infinite set has ultrapowers of arbitrarily large cardinality.

It is natural to ask: When is $P_U A$ κ+-saturated? The next result shows that the answer depends only on the complete theory of A.

Theorem 10.8 (Keisler) Let U be a regular ultrafilter over a set of cardinality κ. If|L| ≤κ and A≡B, then $P_U A$ is κ+-saturated if and only if $P_U B$ is κ+ -saturated.

Let us say that a regular ultrafilter U over a set of cardinality κ saturates a complete theory T if for every model A of T, $P_U A$ is κ+-saturated.

Thus it does not matter which model of T we take. Given two complete theories S, T with countable vocabularies, we write S ⊳ T if every regular ultrafilter which saturates T saturates S. This relation can be used to classify complete theories. Intuitively, higher theories in this ordering are more complex than lower ones. The ⊳-class of T is the set of all S such that S ⊳ T and T ⊳ S. It is clear that ⊳ is is reflexive and transitive, so it induces a partial order on the ⊳-classes. The paper [26] showed that there are at least two ⊳-classes, including a lowest and a highest ⊳-class, and posed several questions which are still open, including: Is this partial order linear? How many ⊳-classes are there? Is there a syntactical characterization of the ⊳-classes? The following theorem gives some partial results.

10.4 SATURATION AND PRESERVATIONS

A type *p*(*x*) is said to be isolated by *φ* if there is a formula *φ*(*x*) with the property that " ψ(x) ∈ p(x), φ(x) → ψ(x). Since finite subsets of a type are always realized in M, there is always an element *b* ∈ *Mn* such that *φ*(*b*) is true in M; i.e. M models *φ*(*b*), theefoe *b* realizes the entire isolated type. So isolated types will be realized in every elementary substructure or extension. Because of this, isolated types can never be omitted.

Consider the language with one binary connective, which we denote as ∈. Let M be the model <ω, ∈ω>, which is the ordinal ω with its standard well-ordering. Let T denote the theory of this model.

Consider the set of formulas p(x) : {n ∈x|n$\in_\omega$ω). First, we claim this is a type. Let p0(x) ⊆ p1(x) . We need to find an n ∈ω that satisfies all the formulas in p0. Well, we can just take the successor of the largest ordinal mentioned in the set of formulas p0(x). Then this will clearly contain all the ordinals mentioned in p0(x).Thus we have that p(x) is a type. Next, note that p(x) is not realized in M. For, if it were there would be some that n $\in \omega$ contains every element of ω. Consider supermodels of M does not always work since the extension models might not be elementary.

To realize the type in an elementary extension new structures have to be defined in this language. More general statements are as follows:

Theorem 10.9 (Morley and Vaught). Let A be an infinite model and let κ be an infinite cardinal such that $\kappa \geq \Sigma_A$. Then A has a κ+-saturated elementary extension of power $\|A\|\kappa$.

Corollary 10.1 Let T be a complete theory of cardinality $\leq \kappa$, and suppose that

T has an infinite model. Then T has a κ+-saturated model of power 2^κ.

Proof. By Löwenheim-Skolem let A be a model of T of power κ. Apply the previous theorem to get a κ+-saturated elementary extension of power $\kappa k = 2^\kappa$.

Corollary 10.2 Assume the Generalized Continuum Hypothesis. Let T be a complete theory of cardinality $\leq \kappa$. Then T has a saturated model of power κ+. This model is unique up to isomorphism.

Proof (Exercise)

The previous corollary may be applied to give the following useful variant of Vaught's test. Unlike Vaught's test itself, the following provides a necessary and sufficient condition for completeness.

Theorem 10.10 Assume $2^{\aleph 0} = \aleph_1$. Let T be a countable complete theory with no finite models. T is complete $\Leftrightarrow$ any two saturated models of T of power $\aleph 1$ are isomorphic.

Proof. Immediate from the previous corollary.

However, the statement that every model has a saturated elementary extension is not provable in ZFC. In fact, this statement is equivalent to the existence of a proper class of cardinals κ such that $\kappa^{<\kappa} = \kappa$. The latter identity implies that either $\kappa = \lambda^+ = 2^\lambda$ for some λ, or κ is weakly inaccessible.

Is G.C.H. needed for the above results. For example, let T be the complete theory of dense linear ordering without end points. Then any

$\aleph 1$-saturated model of T has cardinality $\geq 2\aleph 0$. Thus we do not get satisfactory results unless we assume $2^{\aleph 0} = \aleph 1$.

However, the G.C.H. can be eliminated from most applications of saturated models, by observing that the conclusions are usually absolute. Thus, if we are trying to prove that a certain theory T is complete, we may assume G.C.H., apply the variant of Vaught's test referred to above, and then eliminate G.C.H. by noting that completeness is an absolute property of T.

Actually, there are several ways to avoid the need to assume the G.C.H.

(1)Assume G.C.H. and then eliminate it by absoluteness arguments.

(2)Assume the existence of inaccessible cardinals (some people might con- sider this assumption more reasonable than G.C.H.).

(3)Use "special" models (Morley-Vaught).

(4) Use "recursively saturated" models (Barwise-Schlipf).

We shall simply assume the G.C.H. when needed.

Given a set X, the n-th cumulative power set of X is defined recursively by V0(X) = X, Vn+1(X) = Vn(X) $\cup$ P(Vn(X)).

The superstructure over X is the union of the cumulative power sets and is denoted by V (X),

V (X) = $\cup^{\infty}$ V n (X) . n=0

The superstructure V (X) has a membership relation $\in$ between elements of Vn(X) and Vn+1(X), n = 0,1,2,.... We treat the elements of X as atoms, and always assume that $\varnothing \in /$ X and that no x $\in$ X contains any elements of V (X). We then consider the structure V(X) = $\langle$V(X),$\in\rangle$ whose vocabulary has the single binary relation $\in$.

10.5 FUNCTORIAL MODELS AND DESCRIPTIVE SETS

The author presented in 1997 what we refer to by Descriptive Computation applying generic diagrams. Thus the computing model is definable by a generic diagram with a function set The corresponding terminology (Martin 19770 in set theory refers to sets or topological structure definable in a simple way. We further define H[**κ**] **from** (Barwise 1968) - for any cardinal κ ,let H[κ] be the set of sets x such that the transitive closure of x, TC[x], has cardinality less than κ. For κ regular, L<H[κ]> is the language usually denoted by $L_{\kappa,\omega}$. For a preliminary basis (Kiesler-Knight 2004) might be a newer start. From Barwise (1968) on infintary logic and admissible sets, introduce the infintiatry language L_A where A is a countable admissible sets. We are interested in the sublanguages L_A of the language L (= $L_{inf,\omega}$) which allow finite strings of quantifiers and arbitray conjunction and disjunction. We consider formula j of L to be sets, and the language L_A is L∪A. To insure that L_A is a sensible language, we must require that A satisfy certain closure conditions. An extended language of set theory is applied where in addition to the the only relation symbol e, the language allows additional relation symbols Si, that are n_i-ary. is is the Admisisble sets are defined based on rudimentary transitive sets. The definitions apply a separation principle, know as the

There is a distinction between the relation symbol $\in$ and the meta-symbol $\in$.

Definition 10.6 A set is Descriptive Computable iff it is definable by a G-diagram with computable functions.

Proposition 1 For descriptive computable sets the set H[κ] is definable from a generic diagram by recursion.

Theorem 10.11 (Barwise) Let A be admissible, $A \subseteq H(\omega_1)$.

Proposition 10.6 For A an admissible computable set, A is descriptive computable.

Proof Since A is admissible, from [Barwise 68] applying Kunens implicit definability, Gentzen system completeness for Lω1,ω and Beth definability, A ⊆ H[ω1] can be proved. Thus by the proposition A is descriptive.

10.5.1 ADMISSIBLE FRAGMENTS

Theorem 10.12 For A an admissible computable set, A is descriptive computable.

Proof Since A is admissible, from [Barwise 68] applying Kunens implicit definability, Gentzen system completeness for L1, and Beth definability, A H[1] can be proved. Thus by the proposition A is descriptive. ’

From Barwise (1968) on infintary logic and admissible sets, introduce the infintiatry language L_A where A is a countable admissible sets. We are interested in the sublanguages L_A of the language L (= $L_{inf,\omega}$) which allow finite strings of quantifiers and arbitray conjunction and disjunction. We consider formula j of L to be sets, and the language L_A is L ∪ A. To insure that L_A is a sensible language, we must require that A satisfy certain closure conditions. An extended language of set theory is applied where in addition to the the only relation symbol e, the language allows additional relation symbols Si, that are n_i-ary. is is the Admisisble sets are defined based on rudimentary transitive sets. The definitions apply a separation principle, know as the

There is a distinction between the relation symbol e and the metsymbol ∈.

Defintion 10.7 Δ_0 –speparation, where $\Delta_0 (S_1,\ldots S_k)$ formulas of set theory are defined as a closure as the smallest Y such that

(i) for atomic formualsθ, both θ and $\neg\theta$ are in Y, (ii) Y is closed under conjunction and disjunctions on formulas, (iii) if θ is Y then so are $\forall x\ [x \varepsilon y \rightarrow \theta]$ and □x $[x \varepsilon y \wedge \theta]$, denoted by $\forall x \varepsilon y\ \theta$, and $\exists x \varepsilon y\ \theta$.

(b) The Σ(S1,…,Sk) formuals if set theory form the smallest collection Y closed under (i),(ii), (iii)and

(iv) if θ is in Y, $\exists$ x θ is in Y. (c) The Σ1(S1,…,Sk) formulas of set theory form the smallest collection Y containing the Δ_0 (S1,…Sk) formuals and closedunder (iv).

Call a formula θa Δ_0 formual if is a Δ_0 ( ) formula, i.e. no a additional relation symbols are there.

These classes are important in defining end extensions, e,g. (Feferman-Kriesel 1972), (Nourani 1997).

Definition 10.8 A nonempty transitive set A is rudimentary if A satisfies the following

(a) if a,b e A, then a × b and TC({a}) are ∈ A

(b) (Δ_0–separation) if q is any a Δ_0–formula and y is a variable not free in q, the the following is universally true in A: $\exists y \forall x\ [x \varepsilon y \leftrightarrow x \varepsilon w \wedge \theta]$.

For any formula θ and variable y of set theory $\theta^{(y)}$ is the D_0 –formuals obtained from θ, by relativizing all quantifiers in θ to y.

Definition 10.9 A is admissible if A is rudimentary and satisfies the $\Sigma_{\sim}$-reflection principle: if θ is a S- formula and y is a variable not free in θ, the the following is universally true in A: $\theta \; \Delta \; \exists y$ [y is transitive $\wedge \; \theta^{(y)}$].

Let A be an admisisble set and let L(A) denote the collection of formulas of L<inf,ω> that are in A. Call L<A> an admissile fragment(of L<inf, ω >).

If A is rudimentary and a,b ε A the {a,b} and a ∪ b and a ~ b are in A.

Thus every finite subset of A is an elenet of A. In particular $H[\omega] \subseteq A$. If A is rudimentary, then L_A has the following closure properties.;

(i) if $\varphi \; \varepsilon \; L_A$ then $\neg\varphi \; \varepsilon \; L_A$

(ii) if $\varphi \; \varepsilon \; L_A$ and a εA then $(\forall v_a \; \varphi) \; \varepsilon \; L_A$

(iii) if Γ is a finite subset of L_A then $\wedge\Gamma \varepsilon \; L_A$

(iv) if Γ $L_{A,}$ $\Gamma \varepsilon$ *A then* $\wedge\Gamma\varepsilon$ *A.*

Let A be a transitive set . We say that a set $X \subseteq A$ is S_1 on A if there is a S_1 formula which defines X on <A,∈>.

Theorem 10.13 (Barwise 1968) If A is $S\sim_1$ compact then A is Admissible,

There are completeness, interpolations, definability, and coutable compactness from Barwsie 1968.

Completeness- The set of valid sentences of $L_{A \text{ is}}$ S_1 on A.

Compactness- Let A be countable. If A is admissible then A is $\Sigma\sim_1$ compact.

Theorem10.14 (Barwise 1968) If A is Σ1 compact then A is Admissible,

10.5.2 ADMISSIBLE HULLS

Let M be a structure of finite similarity type, M = (|M |,R1,...,Rl). Regarding the elements of | M | as urelements, let us place ourselves within V|M|. Inside V| M |we form the next
admissible set HYP (M) and call it HYPM the admissible hull of M. Technically, HYP M) and HYP ((|M |,R1,...,Rl)) may differ, e.g. the former may not contain urlements at all. V is the element world.

Let U be any set of the elements called urelements. For ordinals α, we define \\$V_{U,\alpha}$ by \\$V_{U,0}$ = U and $V_{U,\alpha+1} = \wp(V_{U,\alpha})$. $V_{U,\lambda}$ = ∪α < VU,á.

Definition 10.10 A model is admissible iff its universe, functions, and relations are definable with admissible sets. From (Nourani 1997):

Theorem 10.15 (1977) *Admissible* models are obtained by taking a reduct from the admissible hull to the Skolem hull definable by a generic diagram.

Let L be an admissible language defined on the Keisler fragment Lω1,ω. The functor on the category L **ω1, B** might be defined with admissible hom sets.

Thus we can apply generic diagrams to define admissible computable sets and models. Let us define computable functors (Nourani 1996). The functors define generic sets from language strings to form limits and models.

Theorem 10.16 Admissible models are obtained by taking a reduct from the admissible hull to the Skolem hull definableby a generic diagram.

Proof Chapters 4-7.

Definition 10.11 A model is admissible iff its universe, functions, and relations are definable with admissible sets.

Theorem 10.17 There is a generic functor defining an admissible model. Let L be an admissible language [10] defined on the Keisler fragment. The functor on the category $L_{\omega 1,K}$ might be defined with admissible hom sets.

Proof Follows from chapters 4-7, and the precedings.

Theorem 10.18 The generic functors define an L-admissible model provided the functions and relations in the language fragment are L-admissible.
Proof Follows from chapters 4-7, definitions, and the precedings.

Theorem 10.19 There is a generic functor defining an admissible model. Let L be an admissible language defined on the Keisler fragment $L_{\omega 1,\omega}$.

Proof Follows from chapters 4-7, definitions, and the precedings. c.f. (Kiesler-Knight 2004)

10.6 FILTERS, FRAGMENT CONSTRUCTIBLE MODELS, AND SETS

Positive forcing (author 1981) had defined T* to be a theory T augmented with induction schemas on the generic diagram functions that can define a canonical model. Generic diagrams are defined such that there is a proper diagram defined with a specific function set.The function set might be $\Sigma 1$ Skolem functions for the set theory example. Inspired from Keisler's forcing and omitting types exposition on model-theoretic focing, new forcing properties and positive forcing was presented on $L_{\omega 1,K}$ fragments. Forcing instances were proved as MA's realizations, therefore, stating a correspondence between forcing and the axiom. More specifics were restated at the ASL Montreal 2006 on Functorial generic filters. Based on the positive forcing property, let F= $\wp$(T*) be a family of sets on the forcing poset. From the author's 1981 we have the following with more specifics on ASL Montreal.

From the preceding chapters4-7 the positive forcing T* is a F-generic filter.

Basic focing companions were presented that were consequences to the above. We futher have the following that had explicit applications since 1995 on a functorial model theory. Since 1995 author had presented a funtorial model-theory and a set fragment consistency basis has developed. Example applications has been to have positive realizability, i.e. realizability based on positive logic that allowed the author to define Horn Filters (2007). From there we can state.

In axiomatic set theory, the Rasiowa–Sikorski lemma, named after Roman Sikorski and Helena Rasiowa, an important fundamental facts used in the technique of forcing. In the area of forcing, a subset D of a forcing notion $(P, \leq)$ is called dense in P if for any $p \in P$ there is $d \in D$ with $d \leq p$. A filter F in P is called D-generic if $F \cap E \neq \varnothing$ for all $E \in D$.

Now we can state the Rasiowa–Sikorski lemma:

Theorem 10.20 Let $(P, \leq)$ be a poset and $p \in P$. If D is a countable family of dense subsets of P then there exists a D-generic filter F in P such that $p \in F$.

Proof Since D is countable, one can enumerate the dense subsets of P as $D_1, D_2, \ldots$. By assumption, there exists $p \in P$. Then by density, there exists $p_1 \leq p$ with $p_1 \in D_1$. Repeating, one gets $\ldots \leq p_2 \leq p_1 \leq p$ with $pi \in Di$. Then $G = \{ q \in P: \exists\, i, q \geq pi\}$ is a D-generic filter.

The Rasiowa–Sikorski lemma can be viewed is a weaker form of an equivalent to Martin's axiom. More specifically, it is equivalent to MA($2^{À0}$).

For $(P, \geq) = (\mathrm{Func}(X, Y), \subset)$, the poset of partial functions from X to Y, define $Dx = \{s \in P: x \in \mathrm{dom}(s)\}$. If X is countable, the Rasiowa–Sikorski lemma yields a $\{Dx: x \in X\}$-generic filter F and thus a function $\cup F: X \rightarrow Y$.

If D is uncountable, but of cardinality strictly smaller than $2^{\aleph_0}$ and the poset has the countable chain condition, we can instead use Martin's axiom.

The following theorem from the an author's ASL publication (Nourani 2005) was prompted by a comment-question from Dana Scott on Boolean and Heyting models to the author.

Theorem 10.21 There is a Horn dense counterpart to the Rasiowa-Sikorski lemma.

Proof Form the above and chapter 7 there is an generic Horn filters with a Horn-dense base.'

Positive forcing (author 1981) had defined T* to be a theory T augmented with induction schemas on the generic diagram functions that can define a canonical model. Generic diagrams are defined such that there is a proper diagram defined with a specific function set.The function set might be S1 Skolem functions for the set theory example. Inspired from Keisler's forcing and omitting types exposition on model-theoretic focing, new forcing properties and positive forcing was presented on $L_{w1,K}$ fragments. Forcing instances were proved as MA's realizations, therefore, stating a correspondence between forcing and the axiom. More specifics were restated at the ASL Montreal 2006 on functorial generic filters.

Based on the positive forcing property, let F= Ã(T*) be a family of sets on the forcing poset. Since 1995 author had presented a funtorial model-theory and a set fragment consistency basis has developed. Example applications has been to have positive realizability, i.e. realizability based on positive logic that allowed the author to define Horn Filters (2007). From there we can state. An area the author has not explored, while the above's relations to Godel sets were examined at times, for example, (author 1999), is whether there is a Godel constructible set corresponence to the above, where natural forcing companions can be presented based on, for example, inductive closures on Godel operations based ZF definable fragment sets. Are there statements that can be stated on new axioms,for example, Martin's axiom's generlization, i.e.. the proper forcing axiom and Martin's maximum.

10.6.1 FUNCTORS ON \\V AND BOOLEAN MODELS

Considering the descriptive universe \\V and \\VB Boolean modes, e.g. Scott theorem from (Richter 1978) on logical implication for formulas and models for classes of Boolean algebras can be reduced to only the Boolean algebra over {0,1} is applied.

Theorem 10.22 (Richter) If R is the class of Boolean algebras and R' contains only the Boolean algebra over {0,1} then ⊨ for R is equivalent to ⊨ for R'. '

The following proposition based on my Functorial model theory enclosed:

Proposition 10.6 (Author 2005) Statments on \\V and be carried on $\backslash\backslash V_B$ applying generic definable digrams on set models, e.g. Godel operations definable. '

Since the author's 2005-2007, Andreas Blass's brief (Blass 2010) on the ultrafilters produced by forcing with *F-generic* filters might be more illuminating. In the 1980's, Laflamme gave a combinatorial description of filters and proved additional combinatorial properties of the resulting generic ultrafilters. Recently, Mathias-generic with respect to these ultrafilters have played an important role in work of Dorais in reverse mathematics, and Krautzberger has studied the analogs in the world of idempotent ultrafilters.

An overview on Tukey types of ultrafilters on countable sets is as follows: If U and V are ultrafilters, we say that U is Tukey above V, U ≥T V, if there is a function *f*: U → V which maps cofinal subsets of U to cofinal subsets of V. The study of cofinal types of partially ordered sets grew out of the study of Moore–Smith convergence in topology. When restricted to ultrafilters, the Tukey ordering on ultrafilters is actually a generalization of the well-studied Rudin–Keisler ordering with some analogies and some surprising twists. Some progress towards solving a problem of Isbell of whether there is a model of ZFC in which there is only one Tukey type of nonprincipal ultrafilers is on Blass's brief. That work explicated in joint work with S. Todorcevic. Blass states realizability methods for constructive set theories like IZF and CZF fit into the framework of algebraic set theory. This allows us to simultaneously recover known realizability interpretations of Friedman, McCarty and Rathjen.

EXERCISES

1. Prove Theorem 3 There is a Horn dense counterpart to the Rasiowa-Sikorski lemma.
2. From the Ryll-Nardzewski theorem prove that in terms of Boolean algebras, that if B is a Boolean algebra with F(B) infinite (equivalently B is infinite), then B has a nonprincipal ultrafilter.
3. Prove proposition 10.3.
4. (Kiesler 1964) Apply propostion 10.2 to show that if V, W are free ultrafilters over J, K and U = V ×W, then $V <_{RK} U$ and $W <_{RK} U$. (Hint: We have already observed that V ≤RK U. To show $V <_R K$ U, take A=J, let f be the projection from J×K onto J, prove that B[f] =/ B, and apply Proposition 8.4 (iii).
5. Let T be a complete theory of cardinality ≤ κ, and suppose that T has an infinite model. Then T has a κ+-saturated model of power 2^κ.
   Hint: Apply the Löwenheim-Skolem theorem with theorems in the section on saturation.

6. Given a function $f: I \rightarrow J$ and an ultrafilter U over I, define $f[U]=\{Y \subseteq J : f^{-1}(Y) \in U\}$. Prove that f[U] is an ultrafilter over J.
7. A free ultrafilter U over N is minimal if and only if for each $x \in$ UR there is a function g:N→R such that gU =x and g is either constant, strictly increasing, or strictly decreasing.
8. Assume the Generalized Continuum Hypothesis. Let T be a complete theory of cardinality $\leq \kappa$. Then T has a saturated model of power $\kappa+$. This model is unique up to isomorphism.
   Hint: The G.C.H Apply the the uniqueness result for saturated models.
9. Suppose $h : A \rightarrow B$ is a complete embedding. Then for each finite subset S of B there is a $C \prec B$ such that $S \subseteq C$ and $h : A \rightarrow C$ is an ultrapower embedding.
10. Let V be a free ultrafilter over N. Then both $U \times V$ and $V \times U$ are uniform and countably incomplete, and $P_{U\times V} A \sim= \Pi_{V\times U} A \sim= P_V A$ whenever $|A| < \kappa$. Then $U \times V$ and $V \times U$ are both non-regular.
11. (i) Are there any uniform non-regular ultrafilters on an infinite set of cardinality smaller than the first measurable cardinal.
    (ii) A related question is whether a set A can have ultrapowers whose cardinality is not a power of $|A|$.
12. Prove theorems 10.17-10.19

KEYWORDS

- **definite propositional function**
- **Hausdorff space**
- **ramified theory**
- **Stone–Čech compactification**
- **ultrafilter**
- **ultraproduct**

BIBLIOGRAPHY

Abraham H.; Taub (editor) **1961**, *John von Neumann. Collected works, 1,* Pergamon Press, New York.

Adamek, J.; Rosick´y, J.; Vitale, E. M, *Algebraic Theories. A Categorical Introduction To General Algebra;* The Syntax and Semantics of Infinitary Languages.

Adámek, Jiří; Horst Herrlich, George E.; Strecker *Abstract and Concrete Categories*. John Wiley & Sons. **1990,** ISBN 0-471-60922-6.

Addisson, J. "The Theory of Hierarchies," in Proceedings of the International Congress of Logic, Methodology, and Philosophy of Science, Satnford, **1960,** 26–37.

ADJ-79-Thather, J. W.; Wagner, E. G.; Wright, J. B.; "Notes On Algebraic Fundamentals For Theoretical Computer Science," Mathematical Sciences, IBM T. J.; Watson Research Center, Yorktown Heights, N. Y. Reprint from Foundations Computer Science III; part 2, Languages, Logic, Semantics, J. deBakker and J. van Leeuwen, Ed.; Mathematical Center Tract *109,* 1979.

6ADJ-Goguen, J. A.; Thatcher, J. W.; Wagner E. G.; Wright, J. B.; Initial Algebra Semantics and Continuous Algebras, *JACM,* **1977,** *24,* 68–95.

ADJ-Goguen, J. A.; Thatcher, J. W.; Wagner, E. G.; Wright, J. B. "An Introduction to Categories, Algebraic Theories and Algebras," IBM Research Report, RC5369, Yorktown Heights, NY; April **1975**.

ADJ-Goguen, J. A.; Thatcher, J. W.; Wagner, E. G.; Wright, J. B.; "A Junction Between Computer Science and Category Theory," (parts I, II), IBM T. J.; Watson Research Center, Yorktown Heights, NY, Research Report, RC4526, 1973.

ADJ-Goguen, J. A.; Thatcher, J. W.; Wagner, E. G.; Wright, J. B.; "An Introduction to Categories, Algebraic Theories and Algebras," IBM Research Report RC5369, Yorktown Heights, NY, April 1975.

ADJ-Goguen, J. A.; Thatcher, J. W.; Wagner, E. G.; Wright, J. B.;" Initial Algebra Semantics and Continuous Algebras," IBM Research Report RC5701, November 1975, JACM24, **1977,** 68-95.

ADJ-Goguen, J. A.; Thatcher, J. W.; Wagner, E. G.; Wright, J. B.; "A Junction Between Computer Science and Category Theory," (parts I–II), IBM T. J.; Watson Research Center, Yorktown Heights, NY, Research Report, RC4526, 1973.

ADJ-Goguen, J. A.; Thatcher, J. W.; Wagner, E. G.; Wright, J. B.; "A Junction Between Computer Science and Category Theory," (parts I, II), IBM TJ Watson Research Center, Yorktown Heights, NY, Research Report, RC4526, **1973**.

ADJ-Goguen, J. A.; Thatcher, J. W.; Wagner, E. G.; Wright, J. B.; "An Introduction to Categories, Algebraic Theories and Algebras," IBM Research Report, RC5369, Yorktown Heights, NY, April **1975**.

ADJ-Thather, J. W.; Wagner, E. G.; Wright, J. B.; "Notes On Algebraic Fundamentals For Theoretical Computer Science," IBM T. J.; Watson Research Center, Yorktown Heights, NY, Reprint from Foundations Computer Science, III; part *2,* Languages, Logic, Semantics, deBakker, J.; van Leeuwen, J. Ed.; Mathematical Center Tract *109,* **1979**.

ADJ-Thather, J. W.; Wagner, E. G.; Wright, J. B.; "Notes On Algebraic Fundamentals For Theoretical Computer Science," IBM T. J.; Watson Research Center, Yorktown Heights, NY, Reprint from Foundations Computer Science III; part *2,* Languages, Logic, Semantics, J. deBakker and J. van Leeuwen, Ed.; Mathematical Center Tract *109,* 1979.

Alfred Tarski; *Collected papers,* Steven R.; Givant, Ralph N. McKenzie, Ed.; Birkhäuser, Basel. **1986**.

Alfred Tarski; *Logic, semantics, meta-mathematics. Papers from* ***1923,*** *to* ***1938,*** second ed.; Hackett, Indianapolis, translations by Woodger, J. H. **1983**,

Alfred Tarski; Robert L.; Vaught **1957**, *Arithmetical extensions of relational systems, Compositio Mathematica, 13,* 81–102, reprinted in Tarski **1986**, *3,* 653– 674.

Armstrong, M. A.; *Basic Topology*, Springer; 1st edition (May 1, 1997). ISBN 0-387-90839-0.

Artin E.; Schreier, O.; "Algebraische Konstruktion reeller Korper", Hamb. Abh. **1926,** *5,* 85–99.

Artin, M.; Grothedieck, A.; Vedier, J. L.; Therore des Topos et Cohomolgie Etale des Schemas. Lecture Notes in Mathematics, *1(269), 2(270),* Springer Verlag.

Artin, M.; Grothendieck, A.; Verdier, J. L.; *Séminaire de Géométrie Algébrique du Bois Marie (SGA 4)*. Lecture Notes in Mathematics, Springer Verlag, Exposé, I.; **1972,** *269,* 27.

Artin, Michael; Mazur, Barry *Etale homotopy*, Lecture Notes in Mathematics, No. 100, Berlin, New York: Springer-Verlag, **1969**.

Awodey, S.; Warren, A. M.; Homotopy theoretic models of identity types, Math. Proc. of the Cam. Phil. Soc. **2009**.

Barr, M.; Wells, C.; Toposes, Triples and Theories. Springer-Verlag, **1985**.

Barwick, C. On left and right model categories and left and right bousfield localizations, Homology, Homotopy and Applications, **2010,** *2*, 12.

Barwise, J-72, "Syntax and Semantics of Infinitary Languages," Springer-Verlag Lecture Notes in Mathematics, **1968,** *72,* Berlin-Heidelberg-NY.

Barwise, J. "Implicit Definability and Compactness in Infintary Languges," in The Syntax and Semtics of Infintary Languages, Edited by J.; Barwise, SPringer-Verlag LNM, 72, Berlin-Helidelberg, NY.

Barwise, J. "Infinitary Logic and Admissible Sets," JSL, *34,* 226-252. 1969.

Barwise, J. "Syntax and Semantics of Infintiary Languages," Springer-Verlag Lecture Notes in Mathematics, *72,* **1968,** Berlin-Heidelberg-NY.

Barwise, J. Handbook of Mathematical Logic, North-Holland, Second edition; 1978.

Barwise, J.; Handbook of Mathematical Logic, North-Holland, second edition, (Barwise, J., ed.), **1978**.

Barwise, J.; Robinson, A.; "Completing Theories By Forcing," Annals of Mathematical Logic, **1970,** *2(2)*, 119–142.

Batanin, M. A.; Monoidal globular categories as a natural environment for the theory of weak n-categories, Adv. Math. **1998,** *136(1),* 39–103. Bau89. Hans Joachim Baues, Algebraic homotopy, Cambridge Studies in Advanced Mathematics, *15,* Cambridge University Press, Cambridge, 1989.

Benci, V.; A construction of a nonstandard universe, 471–478 in Advances in Dynamical Systems and Quantum Physics, Albeverio, S., et al., Ed.; World Scientific **1995**.

Benci, V.; Di Nasso, M. Alpha-theory: an elementary axiomatics for nonstandard analysis, Expositiones Math. **2003,** *21*, 355–386.

Benedikt, M. "Ultrafilters which extend measures," J. Symb. Logic **1998,** *63,* 638–662.

Bishop, E.; Foundations of Constructive Analysis, McGraw Hill, New York, Brouwer, L. E. J.; Intuitionitische Zerlegung mathematischer Grundbegriffe, Jber. Deutsch. Math. Verein, **1967**, *33,* 251–256.

Blass, A. "A model-theoretic view of some special ultrafilters," 79–90 in Logic Colloquium 77, Macintyre A., et al., Ed.; North-Holland **1978**.

Blass, A. "The Rudin-Keisler ordering of P-points," Trans. Amer. Math. Soc. **1973,** *179,* 145–166.

Blass, A.; F-generic ultrafilters, Mathematics Department, University of Michigan, **2010,** North American Annual Meeting in Washington, DC.

Boone, W.; "The Word Problem," Annals of Mathematics, **1959**, 207–265.

Booth, D, "Ultrafilters on a countable set," Annals of Math. Logic **1971,** *2,* 1–24.

Borceux, F.; *Handbook of Categorical Algebra 3,* In *Encyclopedia of Mathematics and its Applications*, *53,* Cambridge University Press, 1994.

Borceux, F.; *Handbook of Categorical Algebra: vol 1 Basic category theory* **1994,** Cambridge University Press, (Encyclopedia of Mathematics and its Applications) ISBN 0-521-44178-1.

Bourbaki, Nicolas; *Elements of Mathematics: General Topology*, Addison-Wesley, **1966**.

Bredon, Glen E.; *Topology and Geometry* (Graduate Texts in Mathematics), Springer; 1st edition (October 17, 1997). ISBN 0-387-97926-3.

Brown, K. S.; "Abstract homotopy theory and generalized sheaf coho mology, Transactions of the American Mathematical Society **1973,** *186*, 419–458.

Burris, Stanley N.; Sankappanavar, H. P.; **1981,** *A Course in Universal Algebra.* Springer-Verlag. ISBN 3-540-90578-2.

Caramello, "A topos-theoretic approach to Stone-type dualities" Cambridge Univeristy, Mathematics. Report 2012.

Carnap, R.; Bachman, F.; **1936,** Uber Extremalaxiome, Erkenntnis, **1929**, *6,* 166–188.

Čech, Eduard; *Point Sets*, Academic Press, **1969**.

Chang C. C.; Keisler, H. J.; Model Theory, Third Edition, North Holland Elsevier, **1990**.

Chang, C. C. "Ultraproducts and other methods of constructing models," 85–121 in Sets, Models and Reecursion Theory, Crossley, J. N. Ed.; North-Holland **1967**.

Chang, C. C.; Keisler, H. J.; Applications of ultraproducts of pairs of cardinals to the theory of models, Pacific. J. Math. **1962,** *12,* 835–845.

Chang, C. C.; Keisler, H.; Jerome, Model Theory, Studies in Logic and the Foundations of Mathematics, *73,* North Holland, **1977,** 2nd edition.

Colin McLarty, *Elementary Categories, Elementary Toposes*. Oxford Univ. Press. A nice introduction to the basics of category theory, topos theory, and topos logic. Assumes very few prerequisites, **1992**.

Comfort W. W.; Negrepontis, S.; The Theory of Ultrafilters, Springer-Verlag, **1974**.

Dana S.; Scott *Measurable cardinals and constructible sets, Bulletin de l'Académie Polonaise des Sciences, Série des Sciences Mathématiques, Astronomiques et Physiques,* **1961,** *9,* 521–524.

Dana S.; Scott, *Definitions by abstraction in axiomatic set theory, Bulletin of the American Mathematical Society, 61,* 442.

David MacIver, *Filters in Analysis and Topology* **2004** *(Provides an introductory review of filters in topology and in metric spaces.).*

Dirks, M, The Stone Representation Theorem For Boolean Algebras, August **2011,** U-Chicago, Mathematics Technical Report.

Donder, H.-D.; Regularity of ultrafilters and the core model, Israel J. Math. **1988,** *63,* 289–322.

Ehrig, H. "Embedding Theorems in the Algebraic Theory of Graph Grammars." FCT **1977,** 245–255.

Eilenberg, S.; MacLane, S, Relations between homology and homotopy groups of spaces Ann. of Math. 46 **1945,** 480–509.

Eilenberg, S.; MacLane, S.; Relations between homology and homotopy groups of spaces. II Ann. of Math. 51 **1950,** 514–533.

Eklof, P.; Ultraproducts for algebraists, 105–138 in Handbook of Mathematical Logic, Barwise, J., Ed.; North-Holland **1977**.

Feferman, S G.; Kreisel-66, Persistent and invariant formulas relatives to theories of higher types, Bull Amer. Math. Soc.; **1966,** *72,* 480–485.

Foreman, M.; Magidor, M.; Shelah, S.; Martin's maximum, saturated ideals, and non-regular ultrafilters. Part, II; Annals of Math. **1988,** *127*, 521–545.

Fragment Consistent Kleene Models, Fragment. Topologies, and Positive Process Algebras. Algebraic Topological Methods In Computer Science (ATMCS) III, Paris, France.

Francis Borceux, *Handbook of Categorical Algebra 3, Categories of Sheaves*, Volume 52 of the *Encyclopedia of Mathematics and its Applications*. Cambridge University Press. 1994.

Francis Borceux; *Handbook of Categorical Algebra 1,* Cambridge University Press. **1994,** ISBN 0-521-44178-1.

Franz-Viktor Kuhlmann, *The model theory of tame valued fields*. Department of Mathematics and Statistics, University of Saskatchewan, Canada.

Frayne, T. E.; Morel, A. C.; Scott, D. S.; Reduced direct products, Fund. Math. **1962,** *51,* 195–228 (Abstract: Notices Amer. Math. Soc. **1958,** *5,* 674).

Fremlin, D. H. *Consequences of Martin's axiom*. Cambridge tracts in mathematics, no. 84. Cambridge: Cambridge University Press. ISBN 0521250919, **1984.**_

Friedlander, Eric M. *Étale homotopy of simplicial schemes*, Annals of Mathematics Studies, *104,* **1982,** Princeton University Press, ISBN 978-0-691-08288-2; 978-0-691-08317-9.

Fulton, William, *Algebraic Topology*, (Graduate Texts in Mathematics), Springer; 1st edition (September *5,* 1997). ISBN 0-387-94327-7.

Functorial Consistency, May **1997,** AMS 927, Milwaukee, Wisconsin, Special Session on Applications of Model Theory to Analysis and Topology, 1997. Abstract number 927-03-29.

Gallager, R. G. **1968,** Information Theory and Reliable Communication, Wiley, SBN 471 29048 3. May 03, 2008, 1984.

Generic functors and generic sets, AMS Spring **1996,** Re: 96T-03-54 is the abstract number.

Ghilardi, S.; *Free Heyting algebras as bi-Heyting algebras*, Math. Rep. Acad. Sci. Canada XVI.; **1992,** *6,* 240–244.

Gierz, G.; Hoffmann, K. H.; Keimel, K.; Lawson, J. D.; Mislove M.; Scott, D. S.; *Continuous Lattices and Domains*, In *Encyclopedia of Mathematics and its Applications*, *93,* Cambridge University Press, 2003.

Goldblatt, R.; Lectures on the hyperreals. An introduction to nonstandard analysis. Springer, **1998**.

Gray, J. W.; ed. Categories in Computer Science and Logic, Contemporary Math. **1989,** *92*, 1–7, Amer. Math. Soc. 2000 Mathematics Subject Classification: 18 C10, 03C65. Amer. Math. Soc.; 1989.

Grothendieck and Verdier. *Théorie des topos et cohomologie étale des schémas* (known as SGA4). New York/Berlin: Springer. (Lecture notes in mathematics, 269–270).

Hart, Klaas Pieter; Nagata, Jun-iti; Vaughan, Jerry E. **2004,** *Encyclopedia of general topology*. Elsevier. 155–156. ISBN 978-0-444-50355-8.

Henkin, L. "On Mahematical Induction," American Mathematical Monthly, **1960,** *67*.

Henkin, L.; "The Completeness of First Order Functional Calculus," Journal of Symbolic Logic, **1949,** *14.*

Hinman, P. G.; Recusrion Theoretic Hierarchies, Spring-Verlag, 1980.

Huber, P. J.; Homotopical cohomology and Čech cohomology, *Mathematische Annalen* **1961,** *144,* 73–76.

Huet, G.; Oppen, D. C.; Equations and Rewrite Ruels: A Survey, Book, R. V.; editor. Formal Language Theory: Perspective and Open Problems, Academic Press, NY, 1982.

Hugh Woodin W.; *The axiom of determinacy, forcing axioms, and the nonstationary ideal,* De-Gruyter Series in Logic and Its Applications, 1, **1999.**

Hyting, A.; Intuitionism An Introduction, North Holland, 1956.

Jacob M.; Plotkin (editor), *Hausdorff on ordered sets,* American Mathematical Society, Providence. 2005.

Mathematics 141. North Holland, Elsevier. ISBN 0-444-50170-3.

Jacobs, Bart, *Categorical Logic and Type Theory*. Studies in Logic and the Foundations of Mathematics **1999,** 141. North Holland, Elsevier. ISBN 0-444-50170-3.

Jacobson. Basic Algebra II.; Dover. **2009,** ISBN 0-486-47187-X.

Jerome Keisler, H.; Julia F.; Knight: Barwise: infinitary logic and admissible sets. Bulletin of Symbolic Logic **2004,** *10(1),* 4–36

John Lane Bell **2005,** *The Development of Categorical Logic*. Handbook of Philosophical Logic, Volume 12. Springer. Version available online at John Bell's homepage.

John R.; Myhill, Dana S.; Scott *Ordinal definability, Proceedings of sym- posia in pure mathematics* (Dana S.; Scott, editor), Axiomatic Set Theory, American Mathematical Society, Providence, **1971,** *13(1),* 271–278.

John von Neumann, *Zur Einfu □hrung der transfiniten Zahlen, Acta Litterarum ac Scientiarum Regiae Universitatis Hungaricae Francisco-Josephinae* (Szeged), *sectio scientiarum mathematicarum, 1,* 199–208, **1923,** reprinted in **1961**, *1,* 24–33; translated in van Heijenoort **1967,** 346–354.

Johnstone, P. T, "*Sketches of an Elephant: A Topos Theory Compendium*." Oxford Science Publications. **2002**.

Johnstone, P. T. *Stone Spaces*, **1982,** Cambridge University Press, Cambridge, ISBN 0-521-23893-5.

Johnstone, P. T.; **2002,** *Sketches of an Elephant*, part D, vol 2 John, J. L.; *The Development of Categorical Logic*. Handbook of Philosophical Logic, Volume 12. Springer. **2005**.

Johnstone, P. T.; *Topos Theory*, L. M. S.; Monographs no. 10. Academic Press. ISBN 0-12-387850-0. **1997**.

Johnvon Neumann, Überdie Definition durch transfinite Induktion und er wandte *Fragen der allgemeinen Mengenlehre, Mathematische Annalen,* **1928,** *99,* 373–391, reprinted below, **1961,** *1,* 320–338.

Joyal A.; Moerdijk, I.; *Algebraic set theory*, London Mathematical Society Lecture Note Series, *220,* Cambridge University Press, **1995**.

Julia F.; Knight: An Inelastic Model with Indiscernibles. J. Symb. Log. **1978,** *43(2),* 331–334.

Kaiser, L. Term Rewriting Systems - Informatik www.informatik.uni-bremen.de/agbkb/lehre/rbs/texte/ Klop-TR. pdf

Kanovei V.; Reeken, M.; Nonstandard Analysis, Axiomatically, Springer **2004**.

Keiselr, H. J. Fundamentals of Model Theory, Handbook of Mathematical Logic, A2, (Barwise, Ed.), North-Holland. **1978**.

Keisler, H. J; A survey of ultraproducts, in Logic, Methodology and Philosophy of Science, Bar-Hillel, Y., Ed.; North-Holland **1965**, 112–126

Keisler, H. J. Forcing and the Omitting Types Theorem, Studies in Model Theory, Mathematical Association of America, Morley, M. D., Ed.; **1973**, 96–133.

Keisler, H. J.; "Forcing and the Omitting Types Theorem, in Studies in Model Theory, Mathematical Association of America, New York, **1967**.

Keisler, H. J.; "Limit ultrapowers," Trans. Amer. Math. Soc. **1963,** *107*, 383–408.

Keisler, H. J.; "Ultraproducts which are not saturated," J. Symbolic Logic **1967,** *32,* 23–46.

Keisler, H. J. Foundations of Infinitesimal Calculus (online edition) **2007,** available at www.math.wisc.edu/~keisler.

Keisler, H. J.; Good ideals in fields of sets, Annals of Math. 79 **1964,** 338–359.

Keisler, H. J.; Model Theory for Infinitary Logic, North Holland, Amsterdam, 1971.

Keisler, H. J.; On cardinalities of ultraproducts, Bull. Amer. Math. Soc. **1964,** *70,* 644–647.

Keisler, H. J.; Some applications of the theory of models to set theory, in Logic, Methodology and Philosophy of Science, Nagel E., et al., Ed; Stanford Univ. Press **1962**, 80–86.

Keisler, H. J.; Ultraproducts and elementary classes, Koninkl. Ned. Akad. Wetensch. Proc. Ser. A 64 (Indag. Math.) **1961,** *23,* 477–495.

Keisler, H. J.; Ultraproducts and saturated models, Koninkl. Ned. Akad. Wetensch. Proc. Ser. A 67 (Indag. Math.) **1964,** *26,* 178–186.

Keisler, H. J.; Ultraproducts of finite sets, J. Symbolic Logic **1967,** *32,* 47–57.

Keisler, H.; Tarski, J A.; From accessible to inaccessible cardinals, Fund. Math. **1964,** *53,* 225–308.

Ketonen, J.; Nonregular ultrafilters and large cardinals, Trans. Amer. Math. Soc. **1976,** *224,* 61–73.

Knight, J. F.; "Additive Structure in Uncountable Models for a Fixed Completion" Symb. Log. **1983,** *48(3),* 623–628.

Knight, J. F.; "Prime and Atomic Models." J. Symb. Log. **1978,** *43(3),* 385–393.

Knight, J. F.; A Complete L omega-omega Sentence Characterizing N1. J. Symb. Log. **1977,** *42(1),* 59–62.

Knight, J. F.; Algebraic Independence. J. Symb. Log. **1981,** *46(2),* 377–384.

Knight, J. F.; Generic Expansions of Structures. J. Symb. Log. **1973,** *38(4),* 561–570.

Knight, J. F.; Hanf Numbers for Omitting Types Over Particular Theories. J. Symb. Log. **1976,** *41(3),* 583–588.

Knight, J. F.; Michael Stob: Computable Boolean Algebras. J. Symb. Log. **2000,** *65(4),* 1605–1623.

Knight, J. F.; Minimality and Completions of Symb, A. J.; Log. **2001,** *66(3),* 1447–1457.

Knight, J. F.; Nadel, M. E.; Models of Arithmetic and Closed Ideals. J. Symb. Log. **1982,** *47(4),* 833–840.

Knight, J. F.; Omitting Types in Set Theory and Arithmetic. J. Symb. Log. **1976,** *41(1),* 25–32.

Knight, J. F.; Skolem Functions and Elementary Embeddings. J. Symb. Log. **1977,** *42(1),* 94–98.

Knight, J. F.; Types Omitted in Uncountable Models of Arithmetic. J. Symb. Log. **1975,** *40(3),* 317–320.

Knight, J.; "Generic Expansions of Structures," JSL, **1973,** *38,* 561–570.

Knuth, D. E.; Bendix, P. B.; "Simple Word Problems in Universal Algebra," in Computational Problems in Abstract Algebra, Leech, J., Ed.; Pergamon Press, Oxford and New York, **1969**.

Kochen, S. B.; Ultraproducts in the theory of models, Ann. Math. Ser. **1961,** *2(74),* 221–261.

Kock, A.; Reyes. G.; Doctrines in categorical logic, in Handbook of Mathematical Logic, Barwise, J. ed.; North Holland 1977.

Koppelberg, S.; Cardinalities of ultraproducts of finite sets, J. Symbolic Logic **1980,** *45,* 574–584.

Kripke, S. "Transfinite recursion on admissible ordinals," I, II, JSL, **1964**, *29,* 161–162, abstract.

Kripke, S. A. 'Semantical analysis of intuitionistic logic I'. In: Crossley J. N.; Dummett M. A. E.; (eds.): Formal Systems and Recursive Functions. Amsterdam: North- Holland, **1965,** 92–130.

Kunen, K.; Some points in β(N), Math. Proc. Cambridge Philosophical Soc. **1976,** *80(3),* 385–398.

Kunen, K.; Ultrafilters and independent sets, Trans. Amer. Math. Soc. **1972,** *172,* 199–206.

Kunen, Kenneth, *Set Theory: An Introduction to Independence Proofs*. Elsevier. ISBN 0-444-86839-9. **1980.**

Kunen, Kenneth. **1980,** Set Theory: An Introduction to Independence Proofs. North-Holland. ISBN 0-444-85401-0.

Kusraev, A. G.; Samson Semenovich Kutateladze **1999,** *Boolean valued analysis*. Springer. p. 12. ISBN 978-0-7923-5921-0.

Lambek, J.; Scott, O. J.; Introduction to Higher-Order Categorical Logic Cambridge Studies in Advanced Mathematics, 1988.

Lambekt, J. The Mathematics of Sentence Structure, American Mathematical Monthly, *65,* 154–170.

Lascar, D.; Stability in Model Theory, Longman **1987**.

Lawvere, F. W.; Functorial Semantics of Algebraic Theories, Proceedings of the National Academy of Science 50, **1963,** No. 5 (November 1963), 869–872.

Lipschutz, Seymour; *Schaum's Outline of General Topology*, McGraw-Hill; 1st edition (June 1, 1968). ISBN 0-07-037988-2.

Löwenheim, Leopold "On possibilities in the calculus of relatives", *From Frege to Gödel: A Source Book in Mathematical Logic, 1879-1931* (3rd ed.), Cambridge, Massachusetts: Harvard University Press, **1977,** 228–251, ISBN 0-674-32449-8 (*online copy* at Google Books)

Löwenheim, Leopold "Über Möglichkeiten im Relativkalkül", *Mathematische Annalen* **1915,** *76(4)*, 447–470, doi: 10.1007/BF01458217, ISSN 0025-5831.

Mac Lane, S-71, *Categories for The Working Mathematician*, GTM 5, Springer-Verlag, NY Heidelberg Berlin, 1971.

Mac Lane, Saunders, *Categories for the Working Mathematician* 2nd ed. **1998,** Graduate Texts in Mathematics 5. Springer. ISBN 0-387-98403-8.

Macintye, A. "Model Completeness," Handbook Mathematical Logic, Barwise, J. Ed.; **1978,** North-Holland. ISBN: 978-3-540-04242-6.

Macintyre, A. "Martin's Axiom Applied To Algebraically Closed Groups," Math. Scan. **1973,** *32*, 46–56.

Macintyre, A. "Martin's Axiom Applied To Algebraically Closed Groups," Math. Scan. **1973**, *32*, 46–56.

Macintyre, A. "Twenty years of p-adic model theory", In Logic colloquium '84 (Manchester, 1984), edited by Paris J. B.; et al.; Stud. Logic Found. Math. North-Holland, Amsterdam, **1986,** *120,* 121–153.

Makaai, M. "Admissible Sets and Infinitary Logics" Handbook Chapter A7, (Barwise, editor), Studies in Logic and Foundations, **1981,** *90*.

Makkai, M.; Reyes, G.; First Order Categorical Logic, Springer-Verlag Lecture Notes in Mathematics, NY Heidelberg Berlin, **1977,** *611.*

Malliaris, M.; Realization of φ-types and Keisler's order, Annals of Pure and Applied Logic, in press (2014).

Maltsev, Anatoly Ivanovich, "Untersuchungen aus dem Gebiete der mathematischen Logik", *Matematicheskii Sbornik, n.s.* **1936,** *1,* 323–336

Maria Cristina Pedicchio; Walter Tholen, eds. **2004,** *Categorical Foundations: Special Topics in Order, Topology, Algebra, and Sheaf Theory*. Volume 97 of the *Encyclopedia of Mathematics and its Applications*. Cambridge University Press.

Marshall H.; Stone "The Theory of Representations of Boolean Algebras," A monograph available free online.*Transactions of the American Mathematical Society* **1936,** *40,* 37–111.

Martin, D. A. "Descriptive Set Thoery," in Hand Book of Mathematical Logic, Barwise, J.; ed.; North Holland. 1978.

Mary Ellen Rudin, "Martin's Axiom," Handbook Mathematical Logic, J.; Barwise, editor, **1978,** North-Holland. ISBN: 978-3-540-04242-6.

McLane, S.; Categories For the Working Mathematician, GTM, Springer-Verlag, Berlin-NY-Heildeberg, 1971.

Michael Barr and Charles Wells **1985,** *Toposes, Triples and Theories*. Springer Verlag. Corrected online version at http: //www.cwru.edu/artsci/math/wells/pub/ttt.html. More concise than *Sheaves in Geometry and Logic*, but hard on beginners.

Milies, César Polcino; Sehgal, Sudarshan, K. *An introduction to group rings*. Algebras and applications, Volume 1. Springer, **2002,** ISBN 978-1-4020-0238-0.

Mitchell, B. **1965,** *Theory of categories*, Pure and applied mathematics, *17,* Academic Press, ISBN 978-0-124-99250-4, MR 0202787.

Moredij, I.; Palmgren, E.; "Minimal models of Heyting arithmetic, Journal of Symbolic Logic, **1997,** *62,* 1448–1460.

Morley M.; Vaught, R.; Homogeneous universal models, Math. Scand. **1962,** *11*, 37–57.

Munkres, James; *Topology*, Prentice Hall; 2nd edition (December *28,* 1999). ISBN 0-13-181629-2.

Neumann, B. H.; "The Isompophism Problem for Algebraically Closed Groups," in Word Problems, Boone et.al. Ed.; North-Holland, **1973,** 553–562.

Nourani, C. F "Forcing with Universal Sentences and Initial Models, Annual Meeting of the Association for Symbolic Logic, Boston, MA, December **1983,** Journal of Symbolic Logic, **1984**, *49,* 1444.

Nourani, C. F. "A Model Theoretic Approach to Types, Extension, and Implementation," Programming Symposium, April **1980,** Paris, Spriger-Verlag, LNCS.

Nourani, C. F. "Functorial Consistency, May 1997, AMS *927,* Millwaukee, Wisconsin, Abstract number 927-03-29, 1997.

Nourani, C. F. "Functorial Models, Admissible Sets, and Generic Rudimentary Fragments," Summer Logic Colloquium, **1997,** Leeds.

Nourani, C. F. "How Could There Be Models For Nothing and Proofs For Free," February 1993, Brief version International Domains, Logic, and Programming Workshop, May 1995, Darmstadt, Germany.

Nourani, C. F. "Intelligent Languages-A Preliminary Syntactic Theory," **1993,** In: Proceedings of the Mathematical Foundations CS'98 Satellite workshop on Grammar systems. Kelemenová, A., Ed; Silesian University, Faculty of Philosophy and Sciences, Institute of Computer Science, Opava, **1998,** 281–287.

Nourani, C. F. "Filters, Fragment Constructible Models, and Sets," www.ams.org/meetings/sectional/mtgs-2164-1051-03-12.pdf

Nourani, C. F. "Forcing with Universal Sentences and Initial Models, Annual Meeting of the Association for Symbolic Logic, Boston, MA; December 1983, Journal of Symbolic Logic, **1984,** *49*, 1444.

Nourani, C. F. "Fragment Consistency on Functorial Models," (A Preliminary) AMS, San Francisco, April 06 Reference: 1018-18-90.

Nourani, C. F. "Frgament Ultraproduct Models and Projective Sets," European Summer Logic Colloquium, July **2013,** Portugal.

Nourani, C. F. "Functorial Computability and Generic Definable Models, International Congress, Mathematicians, Berlin, August 18–27, **1998**, p. 8. Sect.: 02 Hopf Algebras elib.zib.de/ICM98/B/sched/Short24.ps

Nourani, C. F. "Functorial Fragment Definable Models," February 2000 ASL Annual Meeting, March **2000**.

Nourani, C. F. "Functorial Metamathematics," Maltsev Meeting, Novosibirsk, Russia, November 1998, math.nsc.ru/conference/malmeet/thesis.htm

Nourani, C. F. "Functorial Model Computing," FMCS, UBC Mathematics Department, Vancouver, Canada, June 2005.

Nourani, C. F. "Functorial Model Theory and Generic Fragment Consistency Models," October **1996,** AMS-ASL, San Diego, January 1997.

Nourani, C. F. "Functorial Model Theory, Generic Functors and Sets," January 16, **1995** International Congress, Logic, Methodology, and Philosophy of Science, Florence, Italy, August 1995.

Nourani, C. F. "Functorial Model Theory," www.logic.univie.ac.at/pipermail/abstract-weekly/2003-January/000014.html-37k. www.math.ucla.edu/~asl/bsl/0803/0803-006.ps, 2003.

Nourani, C. F. "Functorial Models and Generic Limit Functors," April 1996. The abstract number is: 918-18-1508. AMS Contributed Paper, January 1997, San Diego.

Nourani, C. F. "Functorial Projective Set Models," Summer Logic Colloquium, Sofia, Bulgaria July 2009.

Nourani, C. F. "Functorial projective set models," ASL Summer Logic Colloquium, Sofia University from July 31 August 5, **2009,** www.math.ucla.edu/~asl/bsl/1601/1601-004.ps.

Nourani, C. F. "Functors and Model Computation with Hasse Diagrams," Topology and Computing, University of Northern Maine, Mini Conference, Maine, April 1997.

Nourani, C. F. "Generic Limit Functorial Models and Toposes," 15th Summer Conference on General Topology and Its Applications July 26–29, at Miami University, Oxford, **2000**.

Nourani, C. F. "Model, Fields, and Factorization," October 2008 ASL, Notre Dame, Indiana. 2009.

Nourani, C. F. "Nonmontonic Model Culmination," November *23,* **1998,** ASL, Winter 1999, UCSD.

Nourani, C. F. "Positive Forcing and Complexity," **1984,** Written at, S. L. K.; Manchester, England, July 1984.

Nourani, C. F. "Positive Infinitary Forcing and Word Problems," 815th Meeting of the AMS; November **1984,** San Diego, Proc. AMS Notices and the Annual Meeting of ASL; Anaheim, CA; January **1985,** Proc. Journal of Symbolic Logic.

Nourani, C. F. "The Connection Between Positive Forcing and Tree Rewriting," Proc. Logics In Computer Science Conference (LICS) and ASL; Stanford University, July **1985,** Proc. Journal of Symbolic Logic.

Nourani, C. F. "Functorial model theory and infinitary language categories, September 1994," Association for Symbolic Logic, San Francisco, January **1995,** see Association for Symbolic Quarterly, Summer **1996,** for recent abstract.

Nourani, C. F. **1984,** "Forcing, Nonmonotonic Logic and Initial Models," Logic Colloquium, Manchester **1984,** Proc.

Nourani, C. F. **1994,** Solution Set For Categories With Initial Models, *ECCT,* July 1994, Tours, France.

Nourani, C. F. **1995**, "Functorial Model Theory, Generic Functors and Sets," January *16,* International Congress, Logic, Methodology, and Philosophy of Science, Florence, Italy, August 1995.

Nourani, C. F. **1997,** "Functorial Consistency," May 1997, AMS 927, Milwaukee, Wisconsin, 1997, Abstract number 927-03-29.

Nourani, C. F. **2000,** "Generic Limit Functorial Models and Toposes," 15th Summer Conference on General Topology and Its Applications July 26–29, at Miami University, Oxford, 2000.

Nourani, C. F. **2005,** "On Certificates and Models," June **2005,** Memo to Stanford TCS.

Nourani, C. F. **2005,** "Positive Categories and Process Algebras," Draft outline August **2005,** Written to a perfunctory Stanford CS project.

Nourani, C. F. **2005,** Fragment Consistent Algebraic Models, July **2006,** Abstract and presentation to the Categories Oktoberfest, U Ottawa, October 2005.

Nourani, C. F. **2005a** «Fragment Consistent Algebras,» July 2005.

Nourani, C. F. **2005b** "Functorial Model Computing," FMCS, UBC Mathematics Department, Vancouver, Canada, June 2005.

Nourani, C. F. **2005c** "Positive Categories and Process Algebras," Draft outline August **2005,** Written to a perfunctory Stanford CS project. Abstract published at Paris France.

Nourani, C. F. **2006,** "A Sound and Complete AI Action Logic, June 2005," Preliminary version Proceedings MFCSIT-06, August 2006, Cork, Ireland, Kozen, D.; **1990,** "On Kleene algebras and closed semirings." In Rovan, B.; Ed.; Mathematical Foundations of Computer Science 1990, volume 452 of Lecture Notes in Computer Science, pages 26–47, Bansk´a Bystrica, **1990,** Springer-Verlag.

Nourani, C. F. **2006,** Functional generic filter ASL, Montreal, May 2006.

Nourani, C. F. **2007,** Functorial models and positive realizability ASL, Florida, March 2007.

Nourani, C. F. **2008,** "Positive Realizability on Horn Filters," European Summer Meeting of the Association For Symbolic Logic, Colloquium '08, Bern, Switzerland, July 3–July 8, 2008.

Nourani, C. F. **2009,** Positive omitting types and fragment consistency, Gödel Society conference, Brazil, 2009?

Nourani, C. F. **2011,** "Fragment Consistent Kleene Models, Fragment Topologies, and Positive Process Algebras," Algebraic Topological Methods In Computer Science (ATMCS) III www.lix.polytechnique.fr/~sanjeevi/atmcs/program.pdf

Nourani, C. F. **2011,** Filters, Fragment Consistent Models, and Preshaves. AMS Meeting #1068, **2011,** Spring Southeastern Section Meeting, Statesboro, Georgia. March 2011.

Nourani, C. F. 96A, "Computable Functors and Generic Model Diagrams—A Preview To The Foundations, December 1996.

Nourani, C. F. 97A, "Functorial Model Theory and Generic Fragment Consistency Models," October 1996, AMS-ASL, San Diego, January 1997.

Nourani, C. F. 97C, "Functorial Models Defined on Initial Ordered Structures—A Preview," March **1997,** Brief presented at FMCS UBC, Canada.

Nourani, C. F. Functorial Consistency, May 1997, AMS927, Millwaukee, Wisconsin, Abstract number 927-03-29, **1997**.

Nourani, C. F. Functorial Model Theory, Generic Functors and Sets," January 16, **1995,** International Congress, Logic, Methodology, and Philosophy of Science, Florence, Italy, August 1995.

Nourani, C. F. Functorial Models and Infinite Language Categories ASL, Wisconsin, Spring **1996,** BSL, *2(4),* December 1996.

Nourani, C. F. Functorial Models, Admissible Sets, and Generic Rudimentary Fragments, March 1997, Summer Logic Colloquium, Leeds, July 1997, BSL, *4(1),* March 1998, www.amsta.leeds.ac.uk/events/logic97/con.html

Nourani, C. F. Infinite Language Categories and Limit Topology: A Preliminary, Mathematical Foundations Computer Science, MFCS, Boulder, CO, 1994.

Nourani, C. F. Positive Realizability on Horn Filters, Summer Logic Colloquium, Bern, Switzerland, July 2007.

Nourani, C. F.; "Diagrams, Possible Worlds, and the Problem of Reasoning in Artificial Intelligence," Logic Colloquium, Padova, Italy, 1988, Proc.in Journal of Symbolic Logic.

Nourani, C. F.; "Forcing With Universal Sentences," **1981,** Proc. ASL, **1983,** *49,* Boston. MA.

Nourani, C. F.; "Functorial Computability and Generic Definable Models," International Congress Mathematicians, Berlin, August 1998.

Nourani, C. F.; "Functorial Consistency," May **1997,** AMS 927, Milwaukee, Wisconsin, Abstract number 927-03-29. 1997.

Nourani, C. F.; "Functorial Model Theory and Generic Fragment Consistency Models, October 1996, AMS- ASL, San Diego, January 1997.

Nourani, C. F.; "Functorial Model Theory and Infinite Language Categories," September 1994, Presented to the ASL; January 1995, San Francisco. ASL Bulletins 1996.

Nourani, C. F.; "Functorial Models and Generic Limit Functors," April 1996, Proc. AMS, San Diego, January 1997.

Nourani, C. F.; "Functors Computing Models On Hasse Diagrams," Toplogical Computing Foundations, Mini Conference Maine, April 1997.

Nourani, C. F.; "Generic Functors, and Generic Sets," AMS, Burlington, August 1995.

Nourani, C. F.; "Infinite Language Categories, Limit Topology, and Categorical Computing", 1994, Brief Abstract, MFCS, Boulder, Colorado, June 1996.

Nourani, C. F.; "On Induction for Types: Models and Proofs," Revised June 1981, Presented at the 2nd Workshop on Theory and Applications of Abstract Data Types," University of Passau, Germany, 1982.

Nourani, C. F.; "On The Power of Positive Thinking on Trees," Brief Abstract Version Appeared at LICS, Stanford, 1985.

Nourani, C. F.; "Slalom Tree Computing," 1994, AI Communications, 9. No. 4, December 1996, IOS Press, Amsterdam.

Nourani, C. F.; "Solution Sets For Categories," AMS, Burlington, Vermont, August *8,* **1995**.

Nourani, C. F.; "The Connection Between Positive Forcing and Tree Rewriting," Proc. Logics In Computer Science Conference (LICS) and ASL; Stanford University, July 1985, Proc. Journal of Symbolic Logic.

Nourani, C. F.; "Admissible Models and Peano Arithmetic," ASL, March 1998, Los Angeles, CA. BSL, *4(2)*, June 1998.

Nourani, C. F.; "Computable functors and generic models diagrams," December 1996, ICM, Berlin. Functorial Computability and Generic Definable Models, International Congress, Mathematicians, Berlin, August 18–27, **1998**, p. 8. Sect.: 02 Hopf Algebras.

Nourani, C. F.; "Computation on Diagrams and Free Skolemization," Summer Logic Colloquium, JSL, July 1993, England.

Nourani, C. F.; "Descriptive Computing," AMS, *919,* April 1977, Memphis, 97T-03-90.

Nourani, C. F.; "Descriptive Computing: The Preliminary Definition," Summer Logic Colloquium, July 1996, San Sebstian Spain. See AMS April 1997, Memphis.

Nourani, C. F.; "Forcing with Universal Sentences and Initial Models," University of Pennsylvania, 1981–1982, Annual Meeting of the Association for Symbolic Logic, December 1983, Proc. in Journal of Symbolic Logic.

Nourani, C. F.; "Fragment Consistent Algebraic Models," Brief Presentation, Categories, Oktoberfest, Mathematics Department, U Ottawa, Canada, October 2005.

Nourani, C. F.; "Frege, Hilbert, Kant, and computation: a preface to discerning epistemic illusion," ASL, New Orleans, January 2007 BSL *2,* September 2007.

Nourani, C. F.; "Functorial Admissible Models," November 26, **1996,** ASL, MIT, Cambridge, March 1997.

Nourani, C. F.; "Functorial Admissible Models," Spring ASL March 97, MIT. BSL, **1997,** *3.*

Nourani, C. F.; "Functorial Computability and Generic Definable Models," International Congress, Mathematicians, Berlin, August 18–27, 1998. Hopf Algebras elib.zib.de/ICM98/B/sched/Short24.ps

Nourani, C. F.; "Functorial Computability And Initial Computable Models," AMS 927 Milwakee, Wisconsin, 1997. Abstract number 97T-68-191. May 1997, *18(4),* 624.

Nourani, C. F.; "Functorial Consistency," AMS *927,* Milwaukee, Wisconsin, Special Session on Applications of Model Theory to Analysis and Topology, Abstract number 927-03-29, May 1997.

Nourani, C. F.; "Functorial Generic Filters," July 2005 ASL, Montreal, 2006.

Nourani, C. F.; "Functorial Model Theory and Generic Fragment Consistency Models," October 1996, AMS-ASL, San Diego, January 1997.

Nourani, C. F.; "Functorial Model Theory and Infinitary Language Categories," Proc. ASL, January 1995, San Francisco. See BSL, 2, *4,* December 1996.

Nourani, C. F.; "Functorial Model Theory, Generic Functors and Sets," January *16,* 1995, International Congress, Logic, Methodology, and Philosophy of Science, Florence, Italy, August 1995.

Nourani, C. F.; "Functorial Models and Generic Limit Functors," April 1996.

Nourani, C. F.; "Functorial Models and Implicit Complexity," www.icm2002.org.cn/B/Schedule_Section01.htm. 2002.

Nourani, C. F.; "Functorial Models and Infinitary Godel Consistency," March *4,* **1999,** Goedel Conference. 5th Barcelona Logic Meeting and 6th Kurt Godel Colloquium.

Nourani, C. F.; "Functorial Models and Infinite Language Categories," ASL, Wisconsin, Spring **1996,** BSL, *2(4),* December 1996.

Nourani, C. F.; "Functorial Models and Positive Realizability," ASL Gainsville, Florida, March 2007.

Nourani, C. F.; "Functorial Models, Admissible Sets, and Generic Rudimentary Fragments," March 1997, Summer Logic Colloquium, Leeds, July 1997, BSL, *4(1),* March 1998. www.amsta.leeds.ac.uk/events/logic97/con.html.

Nourani, C. F.; "Functorial Projective Set Models," Summer Logic Colloquium, Sofia, Bulgaria. July 2009.

Nourani, C. F.; "Functorial String Models," ERLOGOL-2005, Intermediate Problems of Model Theory and Universal Algebra, June 26–July 1, State Technical University/Mathematics Institute, Novosibirsk, Russia. www.nstu.ru/science/conf/erlogol-2005.phtml; www.ams.org/mathcal/info/ 2005_jun26-jul1_novosibirsk.html.

Nourani, C. F.; "Functorial Syntax and Paraconsistent Logics," February 1999, ASL-APA; May 1999, New Orleans.

Nourani, C. F.; "Functors Computing Hasse Diagram Models" BSL, 3, 1997. February 17, 1997, Mini-Conference, Maine, April 1997.

Nourani, C. F.; "Generic Functors and Generic Sets," AMS Spring 96 issue Re: 96T-03-54 is the abstract number. **1996**.

Nourani, C. F.; "Generic Limit Functorial Models and Toposes," TOPO2000, August2000, Oxford, Ohio. atlas-conferences.com/c/a/e/u/04.htm. AMCA: Summer Conference on Topology and its Applications, Oxford, Ohio, 2000.

Nourani, C. F.; "High Fidelity Computing With Abstract Models," Categories and Recursion ASL-Spring, Kansas City, March 15, **1994**.

Nourani, C. F.; "Higher Stratified Consistency and Completeness Proofs," April 2003, SLK 2003, Helsinki, August 14–20, http: //www.math.helsinki.fi/logic/LC2003/abstracts/

Nourani, C. F.; "IFLCs and Grothendeick topology: Preliminaries," February 8, Accepted to PSSL, Glasgow, May 06, 2006.

Nourani, C. F.; "Infinitary Language Categories, Generic Functors, and Functorial Model Theory," ASL www.math.ucla.edu/~asl/bsl/0803/0803-006.ps

Nourani, C. F.; "Positive Forcing: The Unrecorded Specifics," ASL January **1998,** San Antonio, TX.

Nourani, C. F.; "Positive Infinitary Forcing and Word Problems," 815the Meeting of the AMS; November **1984,** San Diego, Proc. AMS Notices; and the Annual Meeting of ASL; Anaheim, CA.; January **1985,** Proc. Journal of Symbolic Logic.

Nourani, C. F.; "Positive Omitting Types and Fragment Consistency Models," XIV Encontro Brasileiro de Lógica Godel Society Conference April 2006.

Nourani, C. F.; "Positive Realizably Morphisms and Tarksi Models," Summer Logic Colloquium, Wroclaw, Poland July 2007.

Nourani, C. F.; "Solution Sets For Categories With Initial Models," March 16, 1994, Extended Abstract, ECCT-94, July 1994, Tours, France.

Nourani, C. F.; "Solution Sets For Categories," AMS, Burlington, August 1995.

Nourani, C. F.; "The Incredible String Models," Brief WOLLIC, June 1996, and Kosice, July 1996.

Nourani, C. F.; "The Solution Set Theorem For Categories With Initial Models," European Category Theory Conference, Tours, France, July 1994.

Nourani, C. F.; "Virtual Tree Computing, Meta-Contextual Logic, and, VR ASL Spring, Seattle, W. A.; March, BSL 8(3), ISSN 1079-8986, 2002.

Nourani, C. F.; "Functorial Model Theory, Generic Functors and Sets," January 16, 1995, International Congress, Logic, Methodology, and Philosophy of Science, Florence, Italy, August 1995.

Nourani, C. F.; "Functorial Consistency," May 1997, AMS 927, Milwaukee, Wisconsin, Special Session on Applications of Model Theory to Analysis and Topology, **1997**. Abstract number 927-03-29.

Nourani, C. F.; **1982,** "Two part paper On Types, Induction, and Inductive Completeness," Second Workshop on Theory and Applications of Data Types, University of Passau, Passau, West Germany.

Nourani, C. F.; Abstract linguistics: A brief overview, in ICML, Mathematical linguistics, Tarragona, Catalunia, Spain, **1996**.

Nourani, C. F.; Category Theory for Computing Science. **1990,** Hemel Hempstead, UK.

Nourani, C. F.; Computability, K.R. Reducibility, March 1998, ASL, Toronto, May 1998, BSL, *4(4),* December 1998.

Nourani, C. F.; Conceptual Mathematics: A First Introduction to Categories. Cambridge University Press, Cambridge, UK, 1997, Reprinted with corrections, 2000.

Nourani, C. F.; Descriptive Computing, Summer Logic Colloquium, San Sebastian, Spain, July **1996,** See AMS April **1997,** Abstracts.

Nourani, C. F.; Elementary Theory of the Category of Sets. In Proceedings of the National Academy of Science, **1964**, *52(6),* 1506–1511.

Nourani, C. F.; Filters, Fragment Consistent Models, and Presheaves 2011 Spring Southeastern Section Meeting, Program by Day AMS Sectional Meeting Program by Day. Current as of Friday, March 4, **2011,** *23,* 34 Georgia Southern University.

Nourani, C. F.; Forcing with Universal Sentences and Initial Models," University of Pennsylvania, 1981–1982, Annual Meeting of the Association for Symbolic Logic, December 1983, Proc. in Journal of Symbolic Logic.

Nourani, C. F.; Fragment Consistency on Functorial Models, AMS, San Francisco, April AMS Reference: 1018-18-90.

Nourani, C. F.; Fragment Consistent Algebraic Models, 2005, Brief Presentation, Categories, Oktoberfest, Mathematics Department, U Ottawa, Canada, October 2005.

Nourani, C. F.; *Functional generic filters ASL*, Montreal, May 2006.

Nourani, C. F.; Functorial Admissible Models, 1996–97 Annual Meeting of the Association for Symbolic Logic, Cambridge, Massachusetts, March 22–25, 1997.

Nourani, C. F.; Functorial Computability and Generic Definable Models, International Congress Mathematicians, Berlin, August 18–27, **1998**.

Nourani, C. F.; Functorial Consistency, May 1997, AMS *927,* Milwaukee, Wisconsin, Special Session on Applications of Model Theory to Analysis and Topology, 1997, Abstract number 927-03-29.

Nourani, C. F.; Functorial Fragment Definable Models, February 2000 ASL Annual Meeting, March 2000, BSL.

Nourani, C. F.; Functorial Generic Filters, July 26, 05 ASL, Montreal, May 06 BSL 2, September 2007.

Nourani, C. F. Functorial generic filters. Annual Meeting of the Association for Symbolic logic-Addendum, Universite duquebec" Montreal Montreal, Quebec, Canada May 17–21, **2006,** Bulletin of Symbolic logic Volume 13, Number 3, Sept. 2007.

Nourani, C. F.; Functorial Hasse Models SLK 2002, European Sumer Logic Colloquium, Munster, Germany, July 2002, wwwmath.uni-muenster.de/LC2002/presentedbytitle.html

Nourani, C. F.; Functorial Model Computing, FMCS, Mathematics Department, University of British Columbia, Vancouver, Canada, June 2005.

Nourani, C. F.; Functorial Model Theory and Generic Fragment Consistency Models, October 1996, AMS-ASL, San Diego, January 1997. www.math.ucla.edu/~asl/bsl/0303/0303-006.ps

Nourani, C. F.; Functorial Model Theory and Infinitary Language Categories, Proc. ASL, January 1995, San Francisco. See BSL, *2(4),* December 1996.

Nourani, C. F.; Functorial Model Theory, Generic Functors and Sets, January 16, International Congress, 1995.

Nourani, C. F.; Functorial Models and Implicit Complexity International Congress of Mathematicians ICM2002, Beijing, China, August 20–28, **2002**.

Nourani, C. F.; Functorial models and implicit complexity. www.icm2002.org.cn/B/Schedule_Section01.htm

Nourani, C. F.; Functorial Models and Implicit Hierarchy Degrees October 2002 Algebra and Discrete Mathematics Under the Influence of Models, 26–31 July, Hattingen-www.esf.org/euresco/03/pc03101.

Nourani, C. F.; Functorial models and positive Realizability ASL Florida, March 2007, Nourani, C. F.; Positive Omitting Types and Fragment Consistency.

Nourani, C. F.; Functorial models and positive realizability. Association For Symbolic Logic, Annual Meeting University of Florida. Gainesville, Florida. March, www.aslonline.org/files/ann07program.ps BSL 2, September 2007.

Nourani, C. F.; Functorial Models, Admissible Sets, and Generic Rudimentary Fragments, Summer Logic Colloquium, Leeds, England, August 1997.

Nourani, C. F.; Functorial Models, Generic Sets and Recursion Urlements, March **2006,** SLK, Netherlands

Nourani, C. F.; Functorial Semantics of Algebraic Theories. In Proceedings of the National Academy of Science **2000,** *50(5),* (November 1963), 869–872.

Nourani, C. F.; Functorial w-Chain Models, **1998,** Holiday Mathematics Symposium, January 1999, La Cruces, NM, Proc. May Jharke, editor.

Nourani, C. F.; Functors Computing Models on Hasse Diagrams, The Preliminary brief, Single-page abstract Announced at Mini Conference on Toplogy and Computing, U Southern Maine, 1998.

Nourani, C. F.; Functors on V. Memo to D.; Scott and A.; Blass. April 2005, and 2011, respectively.

Nourani, C. F.; Generic Limit Functorial Models and Toposes, TOPO2000, August2000, Summer Conference on Topology and its Applications Oxford, Ohio, atlas-conferences.com/c/a/e/u/04.htm AMCA: 2000.

Nourani, C. F.; Idealism, Illusion, and Discovery, International Conference on Logic, Novosibrisk, August 1999.

Nourani, C. F.; Infinitary Multiplayer Games, March 8, 1999, ASL SLK, Utrecht www.folli.uva.nl/CD/1999/library/logic%20colloquium%2099/abstracts/nourani2.pdf Functorial w-Chain Models, 1998, Holiday Mathematics Symposium, La Cruces, NM, Proc. Editor May Jharke. January 1999.

Nourani, C. F.; Introduction to Higher Order Categorical Logic. 1986, Cambridge University Press, Cambridge, UK.

Nourani, C. F.; Moudi, R. M.; "Fields, Certificates, and Models (Preliminary report), AMS, Clairemont, May 2008. Vaught, R. 1958, Prime Models and Saturated Models, Notices American Mathematical Society, 5. 780 Vaught, R 1961, Denumerable Models of Complete Theories, Infinitistic Methods, Pergamon, London, 303–321.

Nourani, C. F.; Positive Omitting Types and Fragment Consistency Models, XIV Encontro Brasileiro de Lógica, Godel Society Conference, April 2006.

Nourani, C. F.; Quantifiers and Sheaves. In Proceedings of the International Congress on Mathematics (Nice 1970), Gauthier-Villars **1971,** 329–334.

Nourani, C. F.; Scott, P. J.; *Introduction to Higher Order Categorical Logic*. Fairly accessible introduction, but somewhat dated. The categorical approach to higher-order logics over polymorphic and dependent types was developed largely after this book was published, **1986**.

Nourani, C. F.; Sets for Mathematics. Cambridge University Press, Cambridge, UK, **2003**.

Nourani, C. F.; Spring ASL March *97,* MIT. BSL, *3*, **1997,** Platek, R.; Foundations of recursion theory, Doctoral dissertation and Supplement, Stanford, **1966**.

Nourani, C. F.; Th. Hoppe **1994,** "GF-Diagrams for Models and Free Proof Trees," Proceedings the Berlin Logic Colloquium, Humboldt University Mathematics, May **1994,** Universitat Potsdam, Germany.

Nourani, C. F.; The Incredible String Models, Brief Versions Accepted at WOLLIC, June **1996,** and Kosice, July 1996.

Nourani, C. F.; Virtual Tree Computing, Meta-Contextual Logic, and VR February 12, **2001,** ASL Spring **2002,** Seattle, WA, March, BSL *8(3),* ISSN 1079-8986.

Nourani, C.; FRM Models with Creative Certificates AMS; May 2008, Santa Clara, CA, **2008.**

Palmgren, F. "Developments in constructive nonstandard analysis, this Bulletin, **1998**, *14*, 233–272.

Paul Halmos; Givant, Steven 1998, *Logic as Algebra*. Dolciani Mathematical Expositions No. 21. The Mathematical Association of America. **1982,** *Stone Spaces*. Cambridge University Press. ISBN 0-521-23893-5.

Pedicchio, Maria Cristina; Tholen, Walter, eds. **2004,** *Categorical foundations. Special topics in order, topology, algebra, and sheaf theory*. Encyclopedia of Mathematics and Its Applications 97. Cambridge: Cambridge University Press. ISBN 0-521-83414-7. Zbl 1034.18001.

Per Martin-Löf, An intuitionstic theory of types, Technical Report, University of Stockholm **1972,** Fabien Morel and Vladimir Voevodsky, A1-homotopy theory of schemes, Inst.

Peter Arndt, Chris Kapulkin Homotopy Theoretic Models of Type Theory, Technical Report, **2012.**

Philip S.; Hirschhorn, Model categories and their localizations, Mathematical Surveys and Monographs, *99,* American Mathematical Society, Provi Hovey, M.; "Model categories, Mathematical Surveys and Monographs, vol. *63,* American Mathematical Society, Providence, Rhode Island, **1999,** HS98.

Pitts, A. M.; Conceptual completeness for first-order intuitionistic logic an application of categorical logic, Annals of Pure and Applied Logic, **1989,** *41*, 33–81.

Platek, R. Foundations of recursion theory, Doctoral Dissertation Supplement, Stanford, **1966,** *66.*

Positive Omitting Types and Fragment Consistency Models XIV Encontro Brasileiro de Lógica XIV Brazilian Logic Conference, Celebrating Kurt Gödel's Centennial (1906–2006) Brazilian Logic Society and Association for Symbolic Logic, April 24–28.

Pratt, V. R. **1990,** "Action Logic and Pure Induction, ", Invited paper, Logics in AI: European Workshop JELIA '90, ed. J. van Eijck, LNCS **1990,** *478,* 97–120, Springer-Verlag, Amsterdam, NL, Sep, **1990,** Also Report No. STAN-CS-90-1343, CS Dept.; Stanford, Nov.

Puritz, C.; "Ultrafilters and standard functions in non-standard arithmetic," Proc. London Math. Soc. **1971,** *22,* 705–733. **1959,** *21,* 439–446.

Ramsey, F. P.; The foundations of mathematics, Proceedings of the London Mathematical Society, **1925,** 25, 338–384.

Robert Goldblatt, **1984,** Topoi, the Categorial Analysis of Logic (Studies in logic and the foundations of mathematics, 98). North-Holland. A good start. Reprinted **2006,** by Dover Publications, and available online at Robert Goldblatt's homepage.

Robinson, A **1965,** Introduction to Model Theory and the Metamathemtics of Algebras, 2nd edition, North Holland, Amsterdam. (first edition 1963). Robinson, A. 1956.

Robinson, A, "Infinite Forcing in Model Theory," Proc. 2nd Scandinavian Logic Symposium, edited by Fenstad, J. (North Hooland) Amsterdam, 317–340.

Roman Murawski, *Undefinability of truth. The problem of priority: Tarski vs Gödel, History and Philosophy of Logic,* **1998,** *19,* 153–160.

Rudin, M. E. "Homogeneity problems in the theory of Cˇech compactifications," Duke Math. Journal **1956,** *23,* 409–419.

Rudin, M. E. "Types of ultrafilters," in Topology Seminar, Wisconsin **1965,** Annals of Math. Studies **1966,** *60,* 147–151

Runde, Volker; *A Taste of Topology (Universitext)*, Springer; 1st edition (July 6, 2005). ISBN 0-387-25790-X.

Rutherford, D.; Edwin **1965,** *Introduction to Lattice Theory*. Oliver and Boyd.

Sacks, Gerald E. *Saturated Model Theory*, Benjamin, W. A.; Reading, Inc.; Mass.; MR0398817. **1972.**

Saunders Mac Lane and Ieke Moerdijk **1992,** *Sheaves in Geometry and Logic: a First Introduction to Topos Theory*. Springer Verlag. More complete, and more difficult to read.

Schiemer, G. "Carnap on Extremal Axioms, Completeness of The Models, and Categoricity," ASL **2012,** *5(4),* December 2012.

Schroder. E. **1895,** "Vorlesungen über die Algebra der Logik (Exakte Logik). Dritter Band: Algebra und Logik der Relative." Teubner, B. G.; Leipzig, 1895.

Scott, D.; "Outline of a Mathematical Theory of Computation," Technical Monograph PRG-2, Oxford University Computing Lab.; **1970,** Proc. 4th Annual Princeton Conference on Information Sciences and Systems, **1970,** 169–176.

Scott, P. J.; **1986,** *Introduction to Higher Order Categorical Logic*, Cambridge University Press, ISBN 0-521-35653-9.

Scott, W. R.; Alegbraically Closed Groups, Proc. American Mathematical Society **1951,** *2,* 118–121.

Seebach, J.; Arthur Jr.; *Counter examples in Topology*, Holt, Rinehart and Winston **1970,** ISBN 0-03-079485-4.

Shelah, S. **2002,** *An Introduction to Mathematical Logic and Type Theory: To Truth Through Proof*, 2nd ed., Kluwer Academic Publishers, ISBN 1-4020-0763-9.

Shelah, S. Saturation of ultrapowers and Keisler's order, Ann. Math. Logic **1972,** *4,* 75–114.

Skolem, T.; Über die Nicht-Charakterisierbarkeit der Zahlenreihe mittels endlich oder abzählbar unendlich vieler Aussagen mit ausschliesslich Zahlenvariablen, Fund. Math. **1934,** *23,* 150–161.

Skolem, Thoralf "Logico-combinatorical investigations in the satisfiability or provabilitiy of mathematical propositions: A simplified proof of a theorem by L.; Löwenheim and generalizations of the theorem", *From Frege to Gödel: A Source Book in Mathematical Logic, 1879–1931* (3rd ed.), Cambridge, Massachusetts: Harvard University Press, **1977,** 252–263, ISBN 0-674-32449-8 (*online copy* at Google Books).

Skolem, Thoralf "Logisch-kombinatorische Untersuchungen über die Erfüllbarkeit oder Beweisbarkeit mathematischer Sätze nebst einem Theoreme über dichte Mengen", *Videnskapsselskapet Skrifter, I.; Matematisk-naturvidenskabelig Klasse* **1920,** *6,* 1–36.

Skolem, Thoralf "Some remarks on axiomatized set theory", *From Frege to Gödel: A Source Book in Mathematical Logic, 1879–1931* (3rd ed.), Cambridge, Massachusetts: Harvard University Press, **1977,** 290–301, ISBN 0-674-32449-8 (*online copy* at Google Books).

Skolem, Thoralf "Über einige Grundlagenfragen der Mathematik", *Skrifter utgitt av det Norske Videnskaps-Akademi i Oslo, I.; Matematisk-naturvidenskabelig Klasse* **1929,** *7,* 1–49.

Steel, J. R. Hodl(R) *is a core model below* Θ, this Bulletin, **1995,** *1,* 75–84.

Stephen Willard, *General Topology*, **1970,** Addison-Wesley Publishing Company, Reading Massachusetts. *(Provides an introductory review of filters in topology.).*

Steve Awodey, Type theory and homotopy, Preprint, Bar 10; Barwick, C; On left and right model categories and left and right bousfield localizations, Homology, Homotopy and Applications, 12, **2010,** *2,* 245–320.

Steven, V. *Topology via Logic*. Cambridge Tracts in Theoretical Computer Science 5. Cambridge: Cambridge University Press. **1989,** p. 66.

Stone, M. H.; "The representation theorem for Boolean algebras," Trans. Amer. Math. Soc. **1936,** *40,* 37–111.

Tarski, A.; Einfuhrung in die mathematische Logik und in die Methodologie der Mathematik. Springer, Vienna, 1937.

Tarski, A.; Mostowski and Robinson, R. M.; Undecidable Theories, North-Holland, 3rd edition, **1971**. [Keisler 67].

Tarski. A. "Some observations on the concepts of ω-consistency and ω-completeness. In Logic, Semantics, and Metamathematics. 278–295.

Terese. **2003,** Term Rewriting Systems - Cambridge University Press www.cambridge.org › ... › Programming languages and applied logic. The abstract number is: 918-18-1508. AMS Contributed Paper, January **1997**.

Thomas Scott Blyth **2005,** *Lattices and ordered algebraic structures*. Springer. p. 151. ISBN 978-1-85233-905-0.

Thoralf Skolem *Einige Bemerkungen zur axiomatischen Begründung der Men- genlehre, Matematikerkongressen i Helsingfors den 4–7 Juli 1922, Den femte skandinaviska matematikerkongressen, Redogörelse, Akademiska-Bokhandeln, Helsinki, reprintedin,* **1970,** 137–152; translated in van Heijenoort 290–301, 217–232. 1967.

Thoralf Skolem, *Logisch-kombinatorische Untersuchungen über die Erfüllbarkeit oder Beweisbarkeit mathematischer Sätze nebst einem Theoreme über dichte Mengen, Videnskaps-selskapets Skrifter, I,* no. *4,* 1–36, reprinted in **1970,** 103–136. Partially translated in van Heijenoort, 252–263, **1967**.

Thoralf Skolem. **1962,** Abstract Set Theory, Notre Dame Mathematical Lecture Notes, no. *8,* Notre Dame.

van den Berg B.; Moerdijk, I.; Aspects of predicative algebraic set theory II: realizability, arXiv: 0801.2305, accepted for publication in Theoretical Computer Science.

Van Den Berg, "Realizability in Algebraic Set Theory." Fachbereich Mathematik, TU Darmstadt, Schlossgartenstrasse *7,* Darmstadt, Germany, 2102.

van den Dries, L.; "Tame Topology and o-Minimal Structures," London Math. Soc. Lecture Note Series 248, Cambridge Univ. Press, Cambridge, **1998**.

van den Dries, L.; Classical Model Theory of Fields. Model Theory, Algebra, and Geometry MSRI Publications, *39,* **2000**.

Volger, H.; On theories which admit initial structures. Preprint: Universität Passau, 1987.

Weaver, G.; Henkin-Keisler Models, Springer **1997,** Department of Mathematics, University of Wisconsin, 480 Lincoln Drive, Madison.

Willard, Stephen, **2004,** *General Topology*. Dover Publications. ISBN 0-486-43479-6.

William Lawvere, F.; Robert Rosebrugh **2003,** *Sets for Mathematics.* Cambridge University Press. Introduces the foundations of mathematics from a categorical perspective.

William Lawvere, F.; Stephen H.; Schanuel *Conceptual Mathematics: A First Introduction to Categories.* Cambridge University Press. An "introduction to categories for computer scientists, logicians, physicists, linguists, etc." **1997** (cited from cover text).

Yankov, V. A. **2001,** "Brouwer lattice", in Hazewinkel, Michiel, *Encyclopedia of Mathematics*, Springer, ISBN 978-1-55608-010-4.

Yiannis N.; Moschovakis, *Descriptive set theory,* North-Holland, Amsterdam. **1980**.

Zermelo, E. Über Grenzzahlenund Mengenbereiche: *NeueUntersuchungen über die Grundlagen der Mengenlehre, Fundamenta Mathematicae,* **1930**, *16,* 29–47.

Zermelo, E. *Untersuchungen über die Grundlagen der Mengenlehre, I.; Mathematische Annalen,* **1967**, *65,* 261–281, translated in van Heijenoort, 199–215.

INDEX

For Product Safety Concerns and Information please contact our EU
representative GPSR@taylorandfrancis.com
Taylor & Francis Verlag GmbH, Kaufingerstraße 24, 80331 München, Germany

www.ingramcontent.com/pod-product-compliance
Ingram Content Group UK Ltd.
Pitfield, Milton Keynes, MK11 3LW, UK
UKHW020156260926
475702UK00047B/214

9781926895925